For Edith and

on their marr

with love from

their cousin Lucy

DIGGING

DIGGING

Lucy Cadogan

Chatto & Windus
LONDON

Published in 1987 by
Chatto & Windus Ltd
30 Bedford Square
London WC1B 3RP

British Library Cataloguing in Publication Data
Cadogan, Lucy
Digging.
I. Title
823'.914 [F] PR6053.A33/

ISBN-0-7011-3240-X

Typeset at The Spartan Press Ltd
Lymington, Hants
Printed in Great Britain by
Redwood Burn Ltd
Trowbridge, Wiltshire

For Gerald

Σκάψε μέσα

George Seferis,
after Marcus Aurelius

AUTHOR'S NOTE

The characters and story of *Digging* are fictional. The transliteration from Greek follows what is familiar rather than what is strictly consistent.

CHAPTER I

The moment the taxi pulled away she remembered she hadn't emptied the coffeepot. Laura groaned, opening and shutting her handbag to see if there was anything else she'd forgotten. They said it was because she didn't empty the pot instantly that she kept having to buy new meshes for the cafetière. Each time it cost her two pounds. What a sorry state this one would be in in two months' time.

Never mind. She was off. Forget the past. Forget her unstripped bed and the vase of wilting dahlias John had sent. Forget John, or try to. London was a beautiful city, and early in the morning early in September London was the perfect place to be leaving. Autumn was ahead, abroad. It was 1971. She was almost forty, too old for romantic walks in an autumn drizzle, and the cross-legged giggles by the gas fire afterwards. She sighed, feeling her years – and suddenly leaned out of the taxi window to wave to a man in a plaid cap walking his grey-nosed labrador along the Embankment. It was so early in the morning that the man waved back.

Even Victoria Station was empty. It was daunting to cross the wide, empty concourse watched by a boy in a baseball cap slouched over a suitcase. But the hush of a sleeping metropolis about to spring into its compulsive rush, the ticket man waiting at his lighted window, excited her.

She started to run, afraid that if she didn't, she'd miss the 5.20. She was off, to Bill Courage's excavation in Crete.

It had been Jenny's idea although Laura didn't know the first thing about archaeology – except that everyone on the team that discovered the tomb of Tutankhamun died. When she'd teased Bill, wondering if he'd taken out insurance against that, at least Jenny sniggered even if Bill left the room. But he'd always found her too frivolous. Thank goodness Jenny was different. After all she was Laura's oldest female friend, and was keen to have her on the dig. Jenny had urged Laura to grab this chance of getting right away from normal life. She'd warned Laura that on an excavation nothing was normal or comfortable, and that as it went on Bill became a kind of maniac, all par for the course.

Jenny also mentioned Bill's old friend and colleague Christopher Bendick who would be there, an old-fashioned bachelor, she'd said, and a bit prissy, but a good archaeologist whom Bill respected.

Jenny amused Laura when she spoke about Bill's profession and professional friends; a solemn expression would fix on her face, head tilted, the corners of her long, beautiful mouth pulled down to a preposterous degree in an effort to look as if she took Bill's archaeology seriously. But Jenny was no archaeologist. Laura knew this, and Jenny knew Laura knew it, which should mean that out in Crete Laura would have a good time laughing with Jenny about the others. And what could be more fun than that, a few good laughs with her old friend? Yes, Laura was excited, very excited, and eager now to get there and begin.

Everyone was to meet at Knossos. Jenny had told her that Bill wanted to have a few days to study the finds from last year before they crossed over to the south coast. Bill and Jenny had driven out, taking with them, according to Jenny, a huge amount of string, notebooks, rulers (people were always losing them), graph paper, tea and salt for their cook who hated Greek salt because it stuck to the fingers. It was mind-boggling to think of Jenny setting up house so far from Warwickshire and her Georgian farmhouse with its wisteria and lantern window. If the village on the south coast of Crete was as primitive as Jenny described, with no electricity or plumbing and only the village oven and two gas rings to cook on, how did she manage?

Laura had let the old man push ahead when they arrived at Heraklion airport. His impatience to be off, as he clutched a cardboard box twice his height, moved her. During the journey he'd shown her a photograph of his grandson which she'd held up to the light to see better. He was pleased, though awkward and restive as he peered out of the window at cloud. He pocketed the photograph and said nothing more.

But the queue for immigration moved at a snail's pace. Laura and the old man had to wait outside on the ramp, watched by a policeman at the door. Another policeman inside fingered his gun and stared. The old man tried to see past the couple in front. From across the customs hall came a shout from a woman who held a little boy. The old man saw and looked back at the policeman. He waited very still for his turn, hugging his box. Laura was surprised that he had not waved back.

But when two officials accused the old man of something after examining his passport, he cried. The policeman with the gun was given orders. He grabbed the old man by the arm, ignoring his tears. Laura did not hear the official shout at her. The official clapped his hand on the counter and she flinched. A white door had just closed behind her friend. She handed over her passport. The official fingered through a card index, giving her suspicious looks. She glanced behind her. There were still at least twenty people and they'd already waited forty-five minutes, visitors like herself, in flowery cotton dresses and sport shirts carrying their duty-free bottle of whisky. She noticed more uniforms standing about staring. Laura shivered. She had forgotten that she was coming to Greece under a dictatorship. Even Crete was not free.

She found a taxi and asked to be taken to the Villa Ariadne at Knossos. As the large Mercedes lurched through the potholes, Laura was thrown about on the back seat like a package. Dusty buildings in semi-ruins, sweaters and the garish colours of plastic bowls that hung in the doorways fell away from Laura's eyes, glazed by heat and lack of sleep.

A stone building faced her after she paid the taxi and was left in peace. She started to lug her case up the stone steps which crossed a moat. How strange that this should have been Sir Arthur Evans's home in Crete, which surely should have been a pretty white villa with a blue door into a courtyard shaded by lemon trees and trellises of vines. The dark green wooden shutters and low ochre walls exuded the atmosphere of a wet day in Middlesex.

'Laura!'

Slowly Laura turned around. Jenny looked worried.

'What's wrong?'

'You've made it. Well done.' Jenny grabbed Laura's suitcase and led her away from the gloomy building. 'This isn't where we stay, except in emergencies.' She walked ahead, back the way the taxi had brought Laura, down the gravel drive, to another, smaller, white building called the taverna. 'Come down and I'll make you a cup of tea.'

Jenny poured water into a mug with a tea bag. 'I'll have to introduce you to everyone when you can face it. And there's more still to come,' she sighed.

'Whenever you want me to face it, I'm ready, darling. I'm all ready to go!' Laura swung one arm into the air, clutching her mug of tea with the other hand. Jenny turned, trying not to show that such enthusiasm was out of place, and made tea for herself.

A deep moan was approaching, and it wasn't an English moan. Hinges whined and a gate latched behind someone who shuffled across the yard outside the kitchen. Laura stepped back to let a thick-set woman in slippers come in. In halting Greek, Jenny introduced Laura, who stretched out her hand and gave the wrinkled face a courteous smile. The Greek woman kept hold of Laura's hand while she poured out in a monotone an explanation that was so particularly addressed to Laura that Laura imagined she understood. Trapped by the hand, Laura had to wait until the woman let her go and Jenny could lead them away.

'Whew. I hope everyone isn't quite that intense. My smiles might not last out the course!'

Jenny showed Laura a square room with a blue linoleum floor, and indicated the fourth bed against the far wall. 'You're in the girls' dorm,' Jenny ordered. 'I hope you brought earplugs, because the nightclub goes on all night full blast.'

'I thought you weren't in the best of humours. Darling!' Laura reached out to Jenny and hugged her. 'You look so tired. What can I do?'

Jenny pulled away. 'I'm fine. You'd better have a nap.' As she pushed the screen door to she called that the loos and showers were opposite.

Jenny certainly wasn't looking her best. Laura bit her lip, thinking what a shame it was. Her grey hair was bunched at the back in a green elastic band, her flat brown sandals and loose, flowery dress were drab. To think how often they'd laughed at 'academic drab', the unshaven legs and unpowdered, hairy faces that had disfigured their women's college. Laura wondered if Jenny, like Bill, would be different on an excavation. She threw her suitcase down on the bed and crossed to the sink, sloshing her face from the single, cold tap. Perhaps if, as was so obvious, Jenny found Bill's excavations a strain, she should cut her hair. She would suggest that to her at a calmer moment.

Being here, away, was a godsend after the past year of making a fool of herself over John. The tap dripped. Laura unzipped her bag. But how could she forget John and those dying red dahlias, standing in a girls' dorm on brash blue linoleum in a dig house in Crete? The bereft aura of this horrible room made her ache; the loss of John's familiarity was stark in such a soulless room. She must pull herself together. John had said nothing could come of it. The trouble was she'd gone on hoping something would. That was so typical of her. She knew how beautiful she was, and it was her pride and her vanity that provoked her into being so stupid. She was a great deal more intelligent and attractive than his wife. She amused and excited him. Anyone who laid eyes on them thrilled at the vision of such a perfect pair. She noticed it on the faces of people who passed them in the street. In restaurants she and John talked and laughed so animatedly that couples at other tables gave up their own torpid efforts to watch them. They were a picture of heaven. What had she done wrong?

And who were the other women she was to share this room with? Laura glared at the other beds. The bed placed at a right angle to hers had a camera on it, and a brown towel. Laura couldn't remember when she'd last shared a room with other women. At school probably, twenty-five years ago, with Snakey, Beanpole and Lizzie.

In a burst of self-contempt Laura pulled everything from her bag and threw it all over her bed.

A searing screech pierced the silence of the room. From somewhere outside came another, crescendoing horribly before it was cut off. Laura ran out into the yard with a towel wrapped around her. She was so scared that when she spotted a young man on the verandah, leaning vacantly against the iron rail, she attacked him. He laughed. Then another screech made Laura clap her hands to her ears.

'They're warming up,' the young man reproved her. He was laughing at her.

Laura thought this puny blond with small, blue piggy eyes and a peeling nose was conceited and callous. Pulling the towel tight around her bosom, she retorted that it was a pity other people's fright gave him such pleasure and crossed back to her room, banging the screen door behind her.

Now sitting on the bed opposite was a gawky girl with nearly white hair who was tinkering with the camera.

'Oh, hello. I'm Laura Bell.' Laura approached the girl with an outstretched hand. The girl looked up. A turned-up sneer creased her thin face.

'You've just come, I suppose.' She gave Laura's hand a limp shake.

'Who are you?'

The girl smirked. 'Obviously you're not one for the indirect approach. I'm trying to repair my camera for this evening. Since you asked, my name is Ellen Campbell.'

'What's that terrible noise?'

'Could it be microphones, do you think? Tonight is the great celebration for the village, the *paneyyri*, since their church is Ayia Sophia and today, of course, is Ayia Sophia's day, whoever she was.'

Laura fell onto her bed, shattered. Archaeologists' nerves must be much stronger than most people's. She supposed she was exhausted, remembered her tranquillisers and reached for her handbag. As a rule she disapproved of pills, but Jenny had said it was imperative to keep up the spirits and not be moody. She'd brought tranquillisers with that in mind.

It was not long before the door was pushed open again. Snakey? Beanpole? Lizzie? Giggling, Laura twisted round to see who it was.

'What the hell makes anyone want to be an archaeologist is beyond me,' moaned a small, dumpy girl, pulling off her dusty shoes and collapsing onto a bed. 'Do you know what I've been doing for the past three hours?'

'Archaeology,' Ellen put in.

'I've been lugging boxes of skulls from one room to another. And why, I couldn't tell you. Are you Laura Bell, Jenny's friend?'

'I'm not an archaeologist, if that's what you mean.'

'There's tea if you'd like some. There won't be any dinner,' Jenny instructed them suddenly through the screen door. Laura raised herself and pulled on a T-shirt and old jeans.

The unpleasant young man with the peeling nose was slouched over a plate, spreading a heap of red jam on bread. Jenny held a plate of biscuits, tentatively offering them.

'He seems well away with his bread and jam, darling. Shall I pour you some tea?' Laura reached for the pot.

'Adam, you haven't met Laura Bell.'

Laura stretched out her hand which Adam didn't take, wagging a finger to acknowledge that they'd now been properly introduced. But when Laura gave Jenny a conniving wink, Jenny looked away, backing off to let other people through the door.

'Bill!' Laura jumped up to give him a hug when he lumbered into the room in a dusty pair of khaki trousers, streaks of dirt scarring his face. 'You'd think you'd been down the mines!'

'Hello, Laura,' Bill sighed. The first Greek song started to blare from the microphones outside, while Jenny offered the plate of biscuits to everyone and filled the cups. Everyone munched and swallowed tea in silence, Jenny hoping that Laura would do the same.

'Shouldn't you introduce me to everyone?' Laura interrupted.

Jenny put down the teapot. 'This is Laura Bell. Adam you've met.' She pointed to the youth with the peeling nose. 'He's our architect and is leaving early tomorrow morning for the south coast to carry on with the survey.'

'She knows my name,' Ellen Campbell piped up, raising an arm as if she were in a classroom.

Jenny tripped out everyone's names. Laura's eyes slid from face to face, her head leaning more and more to one side. The cheerful complainer from the girls' dorm turned out to be called Susan something; there was a huge man with a mass of brown hair sprouting out of his shirt front sitting directly opposite, who smiled; on the other side of Bill sat a remarkably beautiful girl who, Laura recognised instantly, would provoke every latent niggle of inadequacy in her own make-up as time went on, but whose beauty when first seen could only surprise; and further on down the table on Bill's side was a man as middle-aged as Laura, Jenny and Bill, very brown, his thinning blond hair bleached by the sun. It was Christopher Bendick, Bill's old friend, of whom Jenny had spoken back in England. When Jenny introduced him he looked solemnly back at Laura for a moment before suddenly lifting his chin and smiling.

Both Jenny and Bill looked unhappy. Bill's head was down, his eyes fixed in front of him as he held on to his teacup. He was a big man with a mass of black curls and heavy black eyebrows, but it didn't befit his pink cheeks and pug nose to look so glum. Jenny made stilted conversation with Adam about his journey the next day and asked if he wanted her to get anything for him and if he'd mind being the first to arrive at the site and if his nose hurt.

Tea was over. The scraping of chairs on the cement floor sounded even more like a classroom as everyone carried their dishes out. Laura ran out into the late afternoon, which was throbbing with Greek music.

What a wonderful light. Laura looked past the spindly, leaning trunks of the umbrella pines to a field of golden vines and squat olive trunks rising up to the sky. This wasn't England. The cloudlessness of the pink sky clashing with the gold vines impressed her. The dusty, pebbly ground under the pine trees had not, even in this fainter evening light, lost its arid brilliance. Perhaps one day Laura would long again for the dark browns of English mud, but not yet.

'Who's for a party?' Laura looked around and saw it was Adam.

'You mean what we're listening to?'

'Wanna go to this here *paneyyri*? There'll be plenny to drink.' Adam's American slur provoked Laura to cut her consonants with a pin-point staccato.

'I'm not partial to noise for its own sake, but perhaps. Who else is going?'

Adam settled into a chair and stretched out his legs. People Laura had just met crossed in front of them, wrapped in towels, carrying dishes of soap. Christopher Bendick, she noticed, had already changed into a green sport shirt and rather well-cut summer trousers; he walked past them carrying a book. 'Are you coming?' Adam asked him.

'I don't think so really. I've got work to do.'

Disappointed that he'd said no and disappeared through a door further along the verandah, Laura excused herself and ran back downstairs to find Jenny.

Jenny and Bill were still in the dining room discussing arrangements. They both looked put out when Laura accosted them from the doorway, but she knew them too well, she told herself, not to push on. She pulled out another chair.

'Are you going to this village thing?'

Bill shuddered. Jenny shook her head slowly.

'The gearbox has packed up. I had to drive all the way back from the south coast in second yesterday,' Bill told her.

'We had to arrange rooms. That's why we'd gone over. The nightmare of arranging rooms!' Jenny complained.

'But it's cleaner than last year. They've paved the *plateia* and planted a few oleanders.' Bill leaned back, looking far too big for the chair.

'It's trying to decide, for one thing, who can live with whom. Whatever you do will turn out wrong,' Jenny continued.

'Where am I going?' Laura asked.

'Those two rooms on the sea, Bill?'

Bill scowled. Inconsiderate of him, when Jenny looked so worried and tired, Laura thought.

'I suppose we'll take one of those double rooms by the loo since we're the only married couple.'

'What a responsibility, my dears,' Laura teased.

'What do you mean by responsibility?' Bill bridled.

Laura caught Jenny's eye and made her smile, a smile which even broke into a short laugh. 'She didn't mean anything,' Jenny sighed.

Laura jumped to her feet. 'What about a drink? My duty-free's upstairs.'

'Good idea!' It was the decisive Bill now, and Laura rushed out for her bottle of whisky.

Even on their second whisky Jenny went on whispering, a bowl of ice and jug of water between them on the bare table. The microphones down the street blared out whining music which sounded like a muezzin. 'We couldn't have her back,' Jenny was insisting about a girl who had been on the team the year before. 'She'd never wear enough clothes. You know, no bra, loose shirt slipping off the shoulder and something she tied around her which I guess she called a skirt, split right up the side. The workmen, you can imagine! She gave *everyone* too many ideas. Jack and Adam were panting like dogs! Which so disgusted the others that if anything was planned and Claire was coming, only Adam and Jack would come as well, and then they became so jealous of each other that only Jack *or* Adam would come. It became impossible to arrange dig outings, because of course Claire always wanted to go and was the first to ask if she could.'

'Her notebooks were excellent,' Bill put in.

Jenny traced a blemish on the table with her finger. Bill had made her blush.

'What sort of outings?' Laura asked.

'To sites up in the mountains, for instance.' Bill swallowed the rest of his whisky.

Jenny wasn't so red now.

'So what did you do? Weren't the others being awfully prudish? What did, I've forgotten his name, the chap our age who's so brown, Bill's friend . . . ?' Laura asked her.

'Christopher Bendick.'

Bill guffawed. 'Didn't notice her.'

'She sounds quite a handful all the same. Poor you.' Laura smiled at Jenny. 'That Adam chap, by the way, conspicuously lacks manners. Both times I've met him he's been rude.'

'He doesn't mean to be,' Jenny defended him. 'He really loved Claire. I watched him do little things for her which Jack *never* did, like refill her glass even if she was sitting on the other side of the table. Little touching things.'

'I'm surprised.' Laura tapped the table.

'Making himself ridiculous on her account. Once when she was shivering, he went inside and asked the café owner's wife for a shawl which he put round her without a word.'

'You seem ill at ease with him, though.'

Jenny's chair clattered to the floor as she jumped up suddenly. 'Margaret's waiting at the airport!' She looked striken.

'She'll catch a taxi,' Bill rebuked her.

'I told her we'd meet her,' Jenny moaned.

'How can we, with a broken gearbox?' Bill twitted her. Laura was uncomfortable. The whisky hadn't worked. Alone with Jenny, they would talk, but in the meantime Bill intended to be difficult. Laura was shocked by him. Couldn't he see that Jenny was right to worry? She wanted to be rid of both of them.

'I'll leave you, I think,' she announced, rising. 'Have a bath, why don't you?' she called back from half way up the stairs when they were out of her sight, and she felt free again to consider their well-being.

CHAPTER 2

Bill wanted Christopher to take over the tombs. In the car just after he had collected Christopher from the airport, Bill described his interview with the Ephor. Bill was both indignant and worried, and Christopher felt sorry for him. But Minoan rites of burial had never been Christopher's principal interest. The architecture of the palaces and farmhouses – their use of a light well in the darkest centre of a building, their elaborate patterning of pavements indoors and out, the fine ashlar blocks facing the walls of their civilised courtyards – was Christopher's special interest. His book on Minoan architecture was only half written but he hoped, perhaps in the coming winter, if not too many other things cropped up, to finish the first draft. He'd been working on it for five years and it threatened him with failure if he didn't complete it. Would he ever, if tombs were now to take up his time?

But he couldn't say no. Bill was far too upset, however much Christopher would have rather continued excavating the settlement which was turning up fascinating new evidence of Minoan styles of building. Last year he'd uncovered a threshold stone in a green limestone which was unique, and also a gypsum dado facing the inside wall of a small village house which, as far as one could tell, was of no particular importance otherwise. Who had lived there? It didn't stand apart from the other similarly small village houses. It was as if you were to find a marble hall inside the door of a small terrace house in an English village.

But Christopher was used to doing what he couldn't avoid doing, and doing it as well as he could. Bill was the director. It was for him to say who was to do what. After all, tombs had never interested Bill any more than they'd interested Christopher. It was unfortunate that in the spring someone's tractor wheels had fallen into the *dromos* of a tomb; the Ephor insisted that the site be excavated this season, before news of a cemetery reached the ears of robbers. The day after Christopher arrived, he and Bill had visited the Ephor in his large office at the museum; Christopher agreed that the Ephor was right, wishing only that he hadn't needed to reproach them for their obvious reluctance by throwing so many 'Aren't I right?'s at them. Yes, he was right, and they promised the Ephor that they'd put half the workforce on to the tombs, with Christopher in charge.

How long a *dromos* should he be expecting before they would reach the chamber? The music from the *paneyyri* disjointed the peaceful evening and made it hard to concentrate on Pini's *Die minoische Gräberkunde*. Christopher regretted that his German wasn't better. The labour of looking up words exasperated him. He pulled at the Anglepoise lamp so that it shone closer to the page of the dictionary, and ran his finger in irritable zigzags across the tiny print.

Someone was knocking on the library door. Christopher looked up from the pool of light on the large, dark brown table. A nervous chuckle preceded the foreman, who was bringing Christopher a plate of *moustalevria*. The 'badger', as they affectionately called him, set the plate down on the table and, holding a bottle up to the light, asked Christopher if he'd like a raki. Reluctantly, Christopher allowed the foreman to pour them a glass each. Christopher's Greek was fluent if badly pronounced.

Badger asked why Christopher wasn't at the *paneyyri*; there were kilos of meat, gallons of raki, dancing, everything that Christopher liked. Christopher banged his glass on the table and toasted Badger's health, both men gulping down the powerful liquid in the semi-darkness. Badger had come for a purpose, so Christopher waited while Badger named the different meats everyone was eating at that very moment a hundred yards away. Badger's nervous laugh told Christopher he hadn't searched for him in the library to discuss rabbit and pig's liver.

'What are we excavating this year?' he asked finally, pushing the plate of *moustalevria* to Christopher.

Christopher carved out a spoonful of the flour and grape mush. 'Bill's the director. You should ask him.' Its musty taste filled his mouth.

'But you know,' Badger pressed him.

Christopher took another gulp of raki.

'We're going to find gold. That's what they're saying.'

'Who is?' Christopher asked.

Badger chuckled. 'You have a nose for tombs. They say you're the lucky one.' Badger had made his point. He slammed his glass on the table and drank to Christopher. Then he went.

Christopher had meant to spend the evening reading Pini. Over the last twenty years he'd been to many *paneyyris*. It wasn't that *paneyyris* weren't wonderful events, the day in the year of every village when the women emerged in their best dresses from domestic obscurity and even danced; it was the women enjoying themselves in the open, laughing with their husbands and their husbands' friends, that moved Christopher most. When finally the man in the village who was considered by everyone the best dancer stood up, the *paneyyri* would be at its height, the excitement feverish. The other dancers would sit down and the lyra player would begin a new dance very slowly. Faster and faster, the man's dancing would gradually become a frenzy of flying legs and leaps which would come to an abrupt, heroic end as the old man – if indeed the dancer was old – shuffled back to his seat basking in the loud applause of his fellow villagers.

Christopher switched off the light and locked the library door. He and Bill had thought no one else knew about the tomb . . . He wondered if Bill had gone down to the *paneyyri* in the end. He would tell Bill what Badger had said.

CHAPTER 3

The next morning when Laura hazarded opening an eye, she saw that Ellen Campbell's bed was already made, the brown towel in place.

Susan groaned and pulled her sheet up higher. Laura felt like death herself. That watery drink which she'd been given at the *paneyyri* was lethal; now that she was sitting up, her head whirled and a headache cut between her eyes. 'So what is Adam's fate? Did you tell him?' she asked the unhappy figure in the other bed, absurd bits of the night before coming back to her now. Adam had asked Susan to read his stars, when the music was so loud that no one could hear.

'I want you to tell me mine,' Laura muttered, rubbing the side of her face. Then pain and nausea overcame her. She'd been a fool. She knew what was to be done. She hadn't been married to an alcoholic for nothing. Off to the loo, Laura bolted herself in.

Feeling a little better, she returned to the dorm and swallowed several aspirins, pulling on her jeans and T-shirt. With several coffees inside her and a dry piece of toast she should be right as rain, though never again would she fall foul of that watery stuff. From now on caution: doubtless other pitfalls lay ahead.

Jenny was in the kitchen with a wispy, dark girl counting knives, forks and spoons in a jumble on the table. She wore the same loose flowery dress as yesterday. She stopped her counting at the sight of Laura. 'You look *green*, Laura.' The knives she was holding clattered back into the jumble of the rest. 'Sit down!' She pushed Laura into a chair. 'Now, tea?'

'Nothing that a few cups of coffee won't put right,' sighed Laura, pressing her temples hard to stop the ache.

'Not tea?'

'Coffee.'

'Tea's better for you.'

'Coffee, darling. Several cups of coffee and I'll help you with that, whatever you're doing.' The blur of cutlery in front of Laura would have baffled anyone. She shut her eyes while Jenny put on the kettle and spooned Nescafé into three mugs.

'Bill's taken the car to the garage and the others are up at the Stratigraphical. Where's Susan? It's an awful job carting all those boxes. You haven't met Margaret, our cook. She was waiting for me at the airport just as I thought. Thank goodness I took a taxi down to collect her. She would have been there all night, wouldn't you, Margaret?' Jenny avoided Laura's melodramatically shut eyes and red lipstick, which only accentuated her condition. She should have known better at her age. 'There's your coffee, Laura, in front of you. Margaret and I have to make an inventory. Tomorrow we go and the lorry will arrive at the crack of dawn. By then everything has to be packed. Plates, blankets and sheets, camp beds, the lot – you must decide, Margaret, what you'll need for cooking.' Opening the cupboard door, Jenny started to pull things out, lids clattering to the cement floor. The table on which Laura was leaning shivered under the impact of a heavy frying pan.

Laura excused herself and went into the dining room where it was quieter. From there she heard Jenny complain to Margaret about people who were carried away by local ways and drank too much although they knew there was a great deal to be done the next morning.

Laura returned to the kitchen. She put on the kettle for more coffee and with her back to Jenny asked what she should do.

'The sheets, if you really think you're up to it.'

'Show me where they are.'

Jenny led Laura back into the dining room and opened a cupboard at the far end. 'We must leave at least ten pairs here, so see what we need to buy. This afternoon I go into Heraklion for the big shop.' Jenny returned to the kitchen.

Laura piled the sheets on the table. She counted the blankets and pillowcases as well, although Jenny hadn't mentioned them. And towels. Laura made another pile of those, writing on the back of her cigarette packet how many there were of each.

Rumbling noises started outside. Laura glanced out of the window and saw a grey sky. She went back to the others to see if they'd noticed.

'The boxes!' Jenny exclaimed. 'Quick. They're outside.' Jenny hauled huge cardboard boxes into the dining room. 'They're hard to get, these.' Margaret followed like an over-large puppy, bouncing on the balls of her feet. Several times she bumped into the door. She was a strange girl, with one large eye that seemed to move around in her head more than the other. It gave her a flimsy look. But her figure was excellent, her narrow hips and thin legs, like a boy's, invitingly angular under her brown gingham skirt. Was she to be the new Claire?

Jenny glanced at Laura's piles. 'You shouldn't have done towels.'

Laura put the towels back and started to fill up one of the cardboard boxes with sheets and pillowcases. It had gone very dark. A flash of lightning made the telephone go 'ping'. Laura turned on the lights and asked Jenny if she should go and get the others before the rain came.

A wind had got up. The pines were bent right over, the palm trees, the bougainvillea and the honeysuckle made a frantic rustle as Laura ran up the path past Sir Arthur's house. The steel door into the storehouse was open and just inside the draughtsman Jack, who'd been such a rival to the architect Adam the year before, sat at a drawing table with a black towel round his head. 'It's about to pour!' Laura announced.

'That's why the bugs have been so bad.' The wind ripped the graph paper off Jack's table. Laura helped him retrieve it. Christopher stood at a table covered in sherds, and finally looked up. 'Has Bill come back?'

Laura shook her head. She told him how busy the cook was, choosing what she'd cook with. Christopher asked Laura how she felt.

'Ghastly!'

He laughed. 'I'm not surprised.'

From the other end of the storehouse emerged the remarkably beautiful girl from tea the day before, brushing off the dust, raising her thin thigh to shake out the bottoms of her trousers. Her silky hair was done up in a red scarf; in a pale blue denim shirt and tight jeans, she looked unbelievably lovely. When Laura said hello she smiled, and reported that she'd been carrying boxes of other people's skulls all morning.

The first large drops of rain darkened penny-sized circles of dust outside the door. 'Come on!' Jack shouted, starting to run. But Laura heard Christopher inside locking up and tried to help him. As she struggled with the bolt of the steel door he pushed her aside and told her to run.

Sheets of water fell on to the dry ground, running off it in rivers. The path was awash, Laura's pink shirt and jeans hung on her like lead. Water slapped against the walls as Laura made it to the verandah, panting, her feet squelching in sodden sandals. They were ruined, that was for sure. A shower of water sprayed the face of Jack, who was standing just behind her when Laura tossed back her hair.

'You're soaked to the skin,' he fussed. He was thinking as he dried his face with the back of his hand that a middle-aged woman, however attractive – and he had already noticed how beautiful Laura must have been when younger – running through the rain in jeans and flimsy sandals, was a ridiculous sight. She had looked frighteningly out of control, like a car careering down a hill without a driver. Jack settled on the only chair, hoping Laura would go and change. Smells of lunch drifted up the stairs.

CHAPTER 4

When Christopher saw how hard the rain was coming down he gave up lunch. He wasn't particularly hungry, and welcomed the chance to have the storehouse to himself. He returned to the inside room where he'd spread out the sherds from level five of his last trench last year, Trench D. He glanced round for his old bush jacket, and a few seconds later, surprised at himself, he realised he was wearing it. He pulled a silver flask out of his breast pocket and slowly unscrewed the top, wondering if level five was contaminated. It should be Middle Minoan IA but he'd spotted already two distinctly Late Minoan IB sherds from marine-style stirrup jars which worried him. The brandy burned his throat; he enjoyed a second swallow which alleviated his worry. But it would be imperative that Bill give him time off from the tombs to take down the baulk and see what he'd missed last year.

The rain clattered on the corrugated cement roof. Christopher laughed at the racket, which sounded as if the storehouse were under enemy fire. If Bill were there he would glance nervously around the room for leaks. Christopher took up the second of the L.M. I sherds and stared at it. He'd mentioned to Bill the foreman's hints about the tombs. Bill's only reaction had been to leave breakfast abruptly with an 'oh damn'. He was worried about his car, of course. Even the 'damn' might have referred to *that*!

Christopher put the sherd back and sat down. He was tired; his arms and legs ached. Another nip of brandy might cure that. He stared into space, his blue eyes fixing on nothing as he propped the open flask on his knee. It was a joy to be alone, with only the rain. It wasn't that there was anything wrong with the others. The beautiful Mary Elizabeth from North Carolina was a pleasure, with her clean, well-brushed hair and narrow thighs, an unusual event on an excavation. She didn't fuss either, and got on with her work. Whether she thought anything interesting he hadn't managed to find out, but there was time. Ellen was ugly, and Jack could be tiresome. That was all.

The curious thing about himself sitting idly sipping brandy was his lassitude. He lacked ambition. Probably it seemed strange that he let Bill direct him. Was it so strange? Perhaps because he was rich. And Bill was ambitious. He liked power, although he wished he didn't, which complicated things for him. He liked to hear himself laying down the law, but it also shamed him. He would duck behind a newspaper or bite his thumbnail when an argument started. But he'd butt in in the end, and the agony he'd suffered trying to keep out of it would be pointless. Christopher smiled and blew his nose. He loved Bill. Bill was an amusing and likable man. It also pleased Christopher that Bill never presumed on Christopher's money. Bill had a job at London University. Christopher didn't. Even so Christopher was probably richer, which that young Adam with a pale face and no manners would think 'unfair'. Bill did not think it was unfair, however.

But was he, Christopher, becoming a dried-up old stick? He often felt uninvolved and it worried him sometimes to be such a dispassionate observer. He was charitable in his observations, he hoped. If he had learned to give way to others, he also believed in them. Consideration For Others had been knocked into him at prep school and at home. His mother, a frail, hypochondriacal lady, had had him fetching her shawl and footstool by the time he was four. Nor had he ever minded. Running errands for others and making them comfortable had never threatened his sense of himself. At school, although he was somewhat bullied, the other boys soon let him be. He avoided them and they respected that.

But he'd never married. Was he so unattractive or was he afraid? He was nearly fifty, and he had actually never slept with a woman. If Bill was ashamed of loving power and importance, Christopher was ashamed of this. Wasn't it outlandish that in 1971 there could be a perfectly presentable man aged forty-eight who was not homosexual, who liked the sight and company of women, who wanted to be loved but who still hadn't ever lain naked with a woman and felt himself penetrate her, working rhythmically with her to a glorious climax . . .

Christopher rubbed his eyes. The rain had let up. Now there was silence overhead and he'd drunk too much brandy. Carefully he fitted his grandfather's flask back into its pocket and rose to his feet. Something was wrong. He was shivering. His teeth chattered uncontrollably. He was going to be ill at the worst possible moment. Poor Bill!

The sound of the lorry grinding backwards into the yard woke Christopher in his narrow bed in the single room. He heard Badger's footsteps on the verandah and his low mumble at Bill and Jenny's door that the *fortigo* had come. Undoubtedly cursing the lorry for coming early, '*o Kyrios* Bill' was outside a minute later telling the driver in his English-sounding Greek to take the lorry up to the storehouse first, for the wheelbarrows and *zembilia*. Christopher smiled. The decrepitude of the old village lorry which did this job every year, the men every year shouting at Lefteris the driver not to break down the gate, amused him. It irritated Bill, Christopher thought, that his great enterprise on the south coast should be launched in such a bedraggled way. Only when Bill had drunk a lot of wine could he see the joke of it. Perhaps such vanity was part of ambition.

There were more footsteps. Any minute he'd hear Jenny's shrill warnings about the breakable plates and the right list.

'Here I am. What can I do, Bill?' called a singsong voice. Who was that? Christopher lay still, listening.

'Hold this list,' Bill growled.

'And what would you like me to do with it?'

It was Laura.

'Tick everything.'

'Tick everything? Now?'

'God, no! As they load the lorry, you tick.'

'Ah.'

Laura sounded good-humoured. Bill's nerves didn't frighten her. She was helping him, which was kind. And brave. Christopher wiped his damp forehead, sorry he couldn't be out there. Probably the others were standing around and gawping, but loading the old lorry was complicated. The archaeologists were soon scared off. Lefteris, Badger and Georgos were in charge, and Bill, who rushed about waving lists, was responsible.

'I'll wait here, Bill. You go up and see if they're doing all right at the storehouse. Don't worry.'

Tell Bill not to worry! Christopher sat up, very slowly moving his leg out from under the covers. If he could be sure he wouldn't faint, he would go out on to the verandah to watch. Would that be drawing attention to himself? Bill and Jenny had been so kind to him the evening before, shocked when they discovered he'd been in bed all the afternoon. Although exhausted from shopping, Jenny had made him rice; Bill had gone back into Heraklion to get medicine. Jenny had fussed with his pillows and found him a clean pair of pyjamas, which pleased him. Christopher liked to be fussed over and was moved by Jenny's concern, tears welling up when she brought him the rice and a pot of tea.

He wanted to see how Laura would do with the lists.

Slowly he wrapped himself in his paisley dressing gown, and opened his door and peered out. He was relieved that no one noticed him shuffle to a chair. Jenny had ordered him not to get up that day or the next.

Laura saw Christopher but was reading the lists too self-consciously to wave. She wanted everyone to be impressed by her command of the boxes, camp beds, trestle tables, storm lamps – which hadn't been packed in anything. They'd break if left loose. She ran down to find another box, too busy to speak to Christopher as she passed him on the verandah. Reappearing with a box, she grabbed some of the linen sherd bags which were in a heap beside the trestles and stuffed them between the lamps. She was too excited by her efficiency; she broke one of the glass chimneys. Christopher watched her run past him with it, still not speaking, to exchange it for an unbroken one from the shelf at the bottom of the stairs. He was impressed that she'd found out already where they were hidden behind the door.

Laura watched nervously as the lorry backed into the yard. It seemed full already with rubber baskets and wheelbarrows and red and white ranging rods which jutted out like the capsized masts of sailing boats. The wheelbarrows were pathetically upside down, their rubber wheels turning in the air. Clutching her list, Laura ran round to the other side to find Bill. 'I'm all ready, Bill.'

The workmen dived for the boxes. The driver, who had blue eyes that looked like porcelain in his dark face, shouted, '*Nero, nero, nero!*' as he threw the boxes to Badger. Suddenly, impatiently, he jumped the three feet up on to the tail and took over from the foreman, but Badger refused to join the others. With Bill helping Georgos to hand things up, Laura could not find things on the list quickly enough; she asked the foreman to pull her up so that she could see better and have more time. When Bill shouted this in Greek the foreman waved Laura to give the lists to him.

'No. I must do it,' Laura shouted back. She pulled herself up by the leg of a wheelbarrow, clutching her lists and embarrassing the foreman. He climbed down, giving her his place.

'*Nero, nero, nero,*' chanted the driver, pushing into Laura with boisterous exuberance.

'What's he saying?' Laura asked Bill, who was just below. Their eyes met. Bill threw back his head and laughed as he grappled with the ungainly top of another trestle table.

Hunched up in his chair on the verandah, Christopher watched, feeling sadly out of it. Bill was lucky. Laura had made him laugh. For an uncomfortably intense moment Christopher wished it had been him and he pulled at his dressing gown. She looked vibrant and frail in her muddy white sandals, waving the list like a flag, at the centre of the party. Once she must have been ravishing, her figure was still nearly as good as the American girl's. He liked her high cheekbones and slanted, Chinese-looking eyes, which turned up into her temples with a strange secretiveness. Not round and accusing like his mother's. Christopher thought how he'd have liked to take over with the lists. He'd have made her comfortable in a chair so that she could watch him.

'What on earth is she up to?' Jenny came out so quietly that Christopher jumped. 'Christopher, you ought to be in bed.'

Bill shouted up to Laura, 'Water, water, water,' translating the driver's mad chant. 'He's telling you.'

'Water?' exclaimed Laura. She turned on the blue-eyed driver with arms outspread. '*Vino, vino, vino!*'

'*Nero, nero, nero!*' The driver was delighted.

Laura tripped; Jenny shouted; Laura grabbed at a wheelbarrow but it hadn't been roped to the side and slid forward, pushing Laura off her feet. She fell sideways, and Bill grabbed her head. Her legs doubled under her in the mud.

How was he to know Laura was so weak on her feet? It wasn't something Bill remembered about her. Falling off lorries like a rickety eighty-year-old! He supposed all those failed love affairs and wrecked marriages had had their effect. It wasn't surprising; he'd always thought she'd end up in trouble. She was the sort who did. But why the hell did he have to find out how crazy she'd gone when she was already out here – it was a huge nuisance. Why hadn't she given him a sign earlier, fallen off their wall at Little Compton on one of those long summer evenings when she'd stand there with her drink and stare at the cows? Was this only the beginning? Quite a beginning, affronting the foreman as well because she wanted to show off and take his place.

Bill was fuming in the kitchen, gulping down glass after glass of water to calm himself. Jenny was bandaging Laura's wretched knee up in their bedroom. At any rate the lorry was packed and they could be off as soon as Laura pulled herself together. It was a pity Laura had to spoil things. Already he had to put up with Christopher being ill. It was a beautiful day and they were about to start. Everything should have been perfect. This was a moment he looked forward to, launching off in their cavalcade of two cars and the lorry for another season. Couldn't Laura for once have shown a little caution, a little wisdom, a shred of consideration for *him*? Bill glared at the tangle of morning glories waving against the wall outside, hating the wide-open blueness that gawped at the brilliant sunlight; they looked as giddy and unreliable as Laura.

Why on earth had he invited her? He knew how impatient she was for excitement. All those years ago at Oxford she'd resented him wanting to get on with his work. Her vanity was inexhaustible. Even then Laura seemed to think she was so ravishing that a man should be satisfied if he could make love to her twenty-four hours a day. And she hadn't changed. She was still a desperately demanding, self-centred person. With eyes wide open, knowing exactly how she'd behave on a dig, he'd asked her to join them because he felt sorry for her. It was how people always disappointed her that made her pitiful. He and Jenny both pitied her; often they discussed her, wondering why Laura could never grow up. Jenny had wanted her to come quite as much as he had, although she also knew how much trouble she'd be. They were compromising the dig to help Laura over another romantic disaster.

He must make a plan. Why not hand Laura over to Christopher? He'd be over his flu in a few days. If she fell into a tomb it would serve her right. Let Christopher see what he could do with her. Maybe a Laura was just what Christopher needed. Good old Christopher. He was letting Christopher dig the tombs. If they found gold Christopher would get the glory. And if Laura broke her neck Christopher would be much better at coping. Bill ran back up the stairs, avoiding the bedroom. He slammed the screen door behind him as he shot out into the yard where the lorry waited and the others – Mary Elizabeth, Ellen, Susan, the latest arrival called Edward, even Jack who refused to cross with the rest of them in the car – stood about as if nothing would ever happen. Angered by the sight of them, Bill ordered Jack not to wait but to catch the bus into town since otherwise he might miss his bus to Ierapetra.

CHAPTER 5

Bill and the others would arrive soon. It was after three. Bill would be in a striped shirt with his sleeves rolled up, pretending there was nothing that could surprise him as he strode up and down the street shaking hands. Adam sighed, slumped over a beer at Vasilis's café. He had no choice but to wait for them although he'd much rather have found his own room and avoided the whole scene of the lorry and the excitement and the others waiting like immigrants for Bill to allot them their rooms.

Adam was worn out by a long morning on the site, by the heat, the glare and frustration so needle-sharp that he had felt sick as he struggled with his tapes to correct the tiny mistakes he'd made last year on the site plan. Now he stared dazed at the trodden tomatoes, dusty bits of bread and rotting pomegranates at his feet, so tired that even the filth didn't bother him.

Small, frail and weak . . . small, frail and weak . . . typical . . . small . . . Hell! Adam mumbled 'hell' to himself a few more times, wishing he had an aspirin to dull his headache. All those small lines on his plan still danced in front of him when he shut his eyes. And those damn insults of Susan's had been a torture. Their poison made him even more tired. He knew she was raving drunk when she said it. But, damn it, she'd told him that it was in his stupid stars to hate himself and die a violent death, squinting at him in a know-all way and telling him he was a nincompoop! When he was least expecting an attack she'd begun portentously with 'I remember. You're Aries.' Last year she'd done the same thing to Claire, telling her that her life would always be in a muddle and she'd never find real happiness. It had made Claire miserable. She swore to Adam that she would never come on an excavation again. Why hadn't Bill told Susan to stay at home this year? Susan wanted to hurt people. She lashed out at them with her astrological mumbo jumbo like a witch. She was evil but Bill wouldn't see it. He strode about in his striped shirts with the blind bonhomie of an idiot. Any minute now he'd be at it again, marshalling the troops for the new campaign. Rah, rah, rah.

Adam stretched out his legs and almost fell off his chair, grabbing at the table. A chicken with a bald neck ran clucking out from under him and darted across to some cucumber skins. Adam waved to Vasilis to bring him another beer.

If he could have avoided a terrible scene with his mother and father, as well as with Bill and Jenny, he'd have refused to come back. If something startling had cropped up in his dull life in England to prevent him coming, it would have been wonderful. But nothing did, and his mother and Jenny were great friends. Bill had known him since he was a boy. And Jenny worried about him to such a degree that it was embarrassing sometimes. So here he was. Think, if he were in London now, doing another term at the Architectural Association, he'd be on some dull project like an adventure playground. At any rate out here the work was worth it. Bill's site was extraordinary. In the winter Adam had drawn a reconstruction of the two floors above, which would have jutted out over the courtyard, making it a cool place to sit even in the middle of the day, with magnificent views. How different the Minoans were from their descendants, who were at that moment deafening him with their clanging and hammering, as Vasilis got on with building his ugly new hotel. Ends of iron bars overhead were a new hazard. Some of the others might gash their heads.

How should he behave with the others? *Small, frail and weak . . .* Adam narrowed his eyes at the piece of newly concreted street two yards from his chair where two old crones were limping into sight, their bony hands and rheumy faces sticking out of dusty black shrouds. He didn't want to be small, frail or weak.

But the others frightened him because not one of them was a friend. He hated Jack and he'd probably have to share a room again with that complacent, opinionated, selfish ass. Jack had no conscience; he cared about no one but himself. Nor did he give a damn that Adam disliked him. How on earth, Adam asked himself rather desperately now, rubbing his tired ankles, was he to cope? And how was he to behave with Susan after she'd insulted him? He mustn't let it show that he was upset. Difficult. Practise a detached view. Could he do that? Like Christopher Bendick who certainly never seemed put out by people. How did he do it? See everyone as slightly ridiculous – was that the trick? Keep in view Bill's messy hair and striped shirts, Jenny's grey pigtail and Susan's fat bottom? Mary Elizabeth's clothes were always too ironed-looking and Ellen had a flat chest and thick ankles. The new chap Edward was awfully young, and Jenny's friend Laura was rather silly. He'd barely laid eyes on Margaret, the cook. And Jack. Hell, what was ridiculous about Jack? Women liked all that hair on his chest. Even the way he sat with his knees wide apart gave the impression that he was at ease.

'Are you sure the others are coming today?' Vasilis banged down another bottle of beer. Adam smiled. Like a cat – he'd watched their cat at home – he slightly closed his eyes in what he hoped looked like calm good humour and felt the small grin fix itself into position. Conscious that his red peeling face and pale eyes could not match the dark, rubbery face and black eyes glaring down at him, nevertheless he kept smiling. It was his first try at the detached look. He continued to smile as he assured the impatient café owner in an even, low voice that the others would be arriving any moment now.

'Then why aren't they here?'

Adam chuckled, an avuncular chuckle he thought, and raised his glass, thanking Vasilis for the beer and drinking to his health.

Vasilis shrugged and walked away.

She'd been there only an hour but the village was no longer idyllic. Laura slowly made her way back to her room with the dustpan and brush, aggrieved that her joy at arriving should be so short-lived. It was deeply unfair.

The small white houses climbing up the side of a low mountain had looked, as they approached, like a haven. She'd reproached Jenny for not telling her how beautiful the village was. Everything about this other side of Crete had seemed so wonderfully different, an hour ago. She'd been in raptures about the high banks of reeds blocking views of the sea, which made a storybook icy sound in the wind. The palms and tamarisk trees and the low, jagged hills puncturing the dry, grey landscape which looked as uninhabitable as the moon had thrilled her. She hadn't wanted to talk about Adam or anyone else. She'd refused to listen; such uncompromising landscape was too moving to be dulled by chat. When they finally stopped the car and climbed out the old women sitting in the street in front of peeling blue doors, their tins of basil and geraniums propped against the white walls, looked to be all that they should.

Now she was on her way back to clean her room and put up the camp bed. Bill had left her, less than an hour ago, to 'carry on' – a Bill-ism if ever there was one. He'd justified rushing away with a fatuous 'I must be off.' She'd started to cry. Her excitement disintegrated, leaving her leaning against the wall inside the door. The filthy gloom in this cell made her shudder. The walls were of cracked plaster which had been recently titivated by a thin coat of whitewash, but jagged edges where the plaster had broken off showed through. Where the paint had stuck better there were denser white streaks which looked like fungus. Part of the ceiling too dangled over her head. There was an armless doll, a broken pink baby's pot, and two split buckets on the floor, everything covered in thick dust and cobwebs, with sweet papers and bits of loo paper blown into the corners. Bill had dumped what was to be her bed in the middle of the room, so tightly rolled up and un-bedlike that there was not even the promise of sleep. She noticed several ominous holes where the wall and floor did not meet. Would she step on a rat? She hated rats. She was terrified of rats.

Laura had gone to find Jenny. Outside there was a heap of rubble she had to climb over, despite her aching knee, to reach the street. Down on the right-hand side was the room that Bill called 'Dig Headquarters'. He'd pointed it out as they rushed past with her camp bed and bag. Jenny, a large patch of sweat staining the back of her cotton dress, was hammering nails into a wall at the back. Margaret was beside her arranging spices on a trestle table. Jenny hadn't seen Laura come in and Laura was loath to speak, ashamed. Even Margaret, that wispy creature from the kitchen at Knossos, was setting out cutting boards, a tray of eggs and some tomatoes. Her confidence in this black hole made her even *look* different, wider at the waist, her nose beakier. Or perhaps it was the Camping Gaz lamp which Jenny had just lit: the chiaroscuro of its glare showed only Margaret's nose and cheek and the inside of her right wrist, the shadows of both women huge on the pitted walls.

Jenny had jumped and fumbled with her hammer when Laura spoke. It was obvious she was not pleased, too busy with hanging up the pots and pans and discussing meals with Margaret. When Laura asked if there was a spare chair and how to clean her room, Jenny sighed, handed her the dustpan and brush, and said chairs were scarce.

'Are you unpacked?' Jenny asked. Her faint effort at courtesy did not come easily.

'Not really.'

'Bill showed you your room?'

'He was in a great hurry to be off,' Laura complained.

'But you know where you are?'

'Yes.'

'Good,' said Jenny, climbing back up on her chair. 'I'm glad. It's going to be a simple supper tonight.' Her stained back was again turned to Laura. 'On Monday Margaret and I shall go into Ierapetra and buy for the week. Right, Margaret?' Margaret didn't answer but an accord between the two women excluded Laura; Margaret, who was cracking eggs into a bowl, looked content.

Perhaps Jenny was angry because Laura hadn't wanted to talk in the car. She had been too upset by her fall, and then too excited to talk. If Jenny had realised how excited she was as they approached the village she would have been shocked. Extreme excitement was childish, which was why Laura kept her mouth shut, but maybe Jenny had interpreted Laura's reticence as criticism.

As Laura pushed open the wooden door into her room, which hung crookedly from its hinges like a dead thing, she wondered if she should have stayed at home. She stared at the neat package of the camp bed. This was a far cry from her house in Fulham. If there'd been a vacuum cleaner she'd have had hope. But all she had was a yellow plastic dustpan, and practically no light except what could squeeze under the door.

'You haven't done much of a job. Aren't you going to?'

Laura dropped the camp bed. Susan stood in the doorway with both hands on her hips and a smirk on her face.

'I'm tired.' Laura glared back at her.

'Actually I've come to use your famous loo. What you need is water and a broom. Get a bucket from the *apotheke*.' She smiled. 'Cheer up.'

Tears again started to dribble down Laura's face. Susan rushed off to the loo. But she marched back into the room a few minutes later and set about erecting a bed from the roll of canvas that was lying in the dust. With her head well down to avoid having to remark on Laura's tears, she warned Laura that Jenny wouldn't be pleased if she couldn't pull herself together. 'She's a disappointed Pisces, you know.'

'And what's that supposed to mean?'

'She'll despise you if she thinks you can't cope.'

'And you must keep up with your running sections. That is of paramount importance. Which means not filling them in later when you feel like it. Does everyone understand?'

Margaret had produced two huge omelettes. Adam was too tired to eat, his plate heaped with a fat slice which he hadn't touched. With a grin on his face, although Bill's harangue was not meant to be amusing, Adam leaned back in his chair and stared at the ceiling. Margaret, who was sitting opposite him, watched anxiously, afraid that Adam didn't like omelettes. The grin, which every second grew more and more smirk-like, disconcerted Laura, but she kept quiet.

Bill, his sleeves rolled up to mean business, leaned on his elbows at the end of the table. He admonished, rather than instructed, them on how to do their work, which irritated Laura. Also those striped shirts he always wore annoyed her. It was either lazy or arrogant not to wear anything else. Even at Oxford he never cared enough about other people to surprise them with different clothes. Didn't Jenny mind? Who but Bill would direct an excavation in a small, dirty village on the south coast of Crete dressed like a banker? What did the villagers think?

The omelette was surprisingly good. Laura gave Margaret a smile, yumming at her with her mouth full.

'A running argument is what I want. When you notice some particular feature in your trench, a change of colour in the earth, say, or a lump of something that you can't be sure about, put it down anyway *and say what you think it is*!' Bill sounded shriller and shriller, a scholar turned sergeant major. 'Make mistakes. Write them down. I want to read those mistakes later, those guesses. They're evidence. Your notebooks last year were a disappointment, all purple passages and interpretations *after* the level was cleared. I don't want to know what you think *in the end*. I want to know how you are thinking *as you go down*!'

For some time Jenny had been tapping the table with her serving spoon. 'Bill? When you're finished can I say a brief word about washing up? Margaret cannot be expected . . .'

'I'm not finished,' Bill snapped. Jenny blushed. She nudged the long bit of omelette on the serving plate with the spoon. 'I haven't got to labelling. First, does everyone understand about the notebooks?' With a beady look Bill dared Jenny to interrupt him again. But just then someone in the street outside cleared his throat and spat; the ugly rasp of phlegm being dredged up silenced the room. Then someone sniggered. Bill glanced behind to check that the door was shut.

'What's this about a tomb?' asked Susan in the hush that followed. Bill started to gobble his omelette.

'Because the workmen were telling Vasilis when we arrived that we'd be finding gold this year,' Susan carried on. Bill's mouth was full. He shook his head. Jenny interrupted then to ask if anyone wanted more omelette; she declared that omelettes were Margaret's speciality, which made Jack groan.

'Which workman?' Bill asked.

'Georgos. They all know.'

Suddenly Bill was hammering on the table. 'I don't want tombs talked about! These people will go mad.'

'Can't you trust us?' Laura asked, arching her eyebrows.

Bill slurped down several glasses of wine and lit a cigarette. His talk had been interrupted before he'd even reached labelling. He hadn't got to sweeping and keeping trenches clean. The tomb gave them all the chance to disappear to their rooms before he'd finished. Typical of a tomb. He hated tombs and wished Susan had kept quiet. But Susan could never keep quiet. She was far too pleased with herself.

Bill could hear Jenny behind the partition mumble about so much wasted omelette. He was tired. The day had gone on and still there was their room. Jenny refused to put their room in order. Every time it was like this. She never considered how tired he must be at the end of the long journey over the mountains. He was the one who had to make all the decisions, settle everyone, put up with everyone's grumbles and see that they weren't forgetting their responsibilities. The washing-up could have waited. There was all of tomorrow, Sunday, to organise a rota. Why had she interrupted? She knew how the work preyed on his mind and how disappointed he'd been with everyone's notebooks last year. In England they'd talked about it. He'd rehearsed with her exactly what he'd say right at the beginning so that there could be no doubt in anyone's mind about how he wanted the work done. Was she so rattled that she'd forgotten? Was it Margaret and her damn omelettes? Couldn't anyone think of him?

Laura heard argument and then silence, the pauses drawing out. Camp beds and suitcases scratched against the floor. How could Jenny and Bill sleep in that room? Neither of them had swept it out. There was still the rubbish which Laura supposed they'd kicked into a corner. Were they arguing about her? Had her falling off the lorry made them regret she'd come? Laura felt sorry for Jenny, she looked so out of place with Bill's team, behaving like a hostess rather than passing the plates down and letting everyone help themselves. How could Jenny show what an amusing, clever person she was with that lot? She could be such a darling, so wise and on the ball, when she was at home with her pretty things and pretty garden and home-made bread. At home she was safe, her friends were polite and affectionate, and Jenny when sure of everyone's love and respect was a delight. Laura had never achieved such a home. Laura envied Jenny, and was often afflicted with a mawkish sense of failure when she went to Little Compton. 'A home of one's own is the home of lost causes.' That's what Jenny would say, curling her lip in that funny way of hers to spout, 'Matthew Arnold, preface to *Essays in Criticism*, First Series, about Oxford.' It was a game they played, which Jenny always won, her knowledge of clichés astounding. None of them here would have guessed that about their director's wife.

Laura rolled over, turning her back to the wall. But the canvas under her had no softness. She rolled back on to her back and stared up into the dark, glad she couldn't see the dangling strands of paint and plaster and cobwebs, also glad at that very moment that she wasn't married to the director of the excavation, whoever he might be – even if he weren't Bill but someone glamorous in a bush jacket, looking like Kirk Douglas, with Dirk Bogarde's lethargy. She hoped things would improve tomorrow when Christopher Bendick arrived.

CHAPTER 6

'Forget the damn jam!'

'But there was a run on it this morning. Two jars of strawberry went just like that. We'll have to ration it.' Jenny had pushed past Bill a second ago with her arms full of bread. Now her calm was theatrical as she came back through the gap in the partition. Laura sat opposite Bill, sipping coffee. He frightened her. She didn't think he was safe to ignore. Sunk down onto a chair with his face covered in his hands, he'd suddenly shouted like a madman. Jenny was asking for it.

'Have some coffee. Let me make you a cup,' Laura offered, moving round to Bill's side of the table and resting a hand on his shoulder.

'It's been robbed. Jenny, listen to me a moment.' Jenny had her eyes on a piece of lined yellow paper lying on the table.

'Yes? I must be off in a minute. Margaret and I have all this to buy.' She waved the list at Bill.

'What do you mean robbed?'

'The tomb! When Christopher and I got there it was all dug out, the blocking stone there on its side. Empty. Everything. Gone.'

Jenny stopped waving the list. Her mouth hung open, which kindled Bill's fury. 'You see what I mean about tombs? I *hate* them. Damn tombs. On the first day of digging, the *first* day, *this* has to happen. That farmer must have told. Everyone knew about it. See? Now I have to telephone and report it to the Ephor. And what is he going to say? He'll blame me. He must blame someone and who else is there?'

Jenny and Laura watched Bill lumber off with his hands in his pockets, a cigarette in his mouth. How solitary he looked retreating down the village street. Laura couldn't understand why someone called an Ephor should worry Bill more than the actual loss of the treasures from the tomb. He was to be pitied certainly, but only because he was so upset. Not because of this monster Ephor who surely wouldn't blame him.

Bill was miserable. That this should happen to him seemed unfair. He despised the 'cult' of tombs. A hoard of valuables had never been his dream. It wasn't archaeology, it was treasure hunting. Also the shock that perhaps one of his workmen from Knossos had had something to do with it shrank his hopes for the season ahead. How could Christopher be so sure it was none of their men from Knossos? Why had Badger mentioned it in the car on Saturday on their way over? And Susan said Georgos had boasted at the café that they were going to find gold. What did that mean? Had Christopher told? No, Christopher wouldn't have breathed a word and until Saturday evening, when Susan brought it up, only Christopher knew. Bill had kept it from the others. How did it become known? The Ephor would have warned the farmer in the spring to keep his mouth shut. The farmer would have had his prize money.

Bill walked on towards the police station, dreading the Ephor's reaction. Sideris Koraes would be sitting at his table in the museum thinking about something quite different. There would be a pause after he heard, while he waited for Bill to suggest what he, the Ephor, should do. He'd point out to Bill that it put *him* in an awkward position, not blaming Bill directly but implying that it was Bill's fault that it had been allowed to happen. If Bill had come to Greece earlier, it wouldn't have happened, in other words. Bill knew that the Colonels in Athens were always on the lookout for new heads. Koraes would explain how it would now get around that he, Koraes, allowed foreigners to lose Greece her heritage. He might lose his job and then where would Bill be? 'So,' he'd conclude. That awful 'so'. Bill knew the finds must be recovered. Koraes wouldn't need to tell him that, but nevertheless he'd be sure to ask Bill how he proposed to recover them.

Christopher could be extraordinarily naive. Right now he and Adam were walking that hill, calmly planning four new trenches, as if nothing had happened. Christopher said Bill should trust the Ephor. The very fact that the Ephor had insisted Bill dig the tombs this season meant he knew robbers could find out. They'd found out sooner than everyone would have liked. That's all, said Christopher. That's all! It was not Bill's fault, Christopher had reassured him just now, the ass. God, Christopher sometimes was too much. Too calm and gentlemanly by half. He was so well bred, there was nothing left of him. That's what Bill sometimes quite seriously thought, and was thinking now. 'Tell Koraes I'm putting down at least four trial trenches to see if there are more unrobbed tombs.' As if that would make up for the other.

Up ahead, where the orange groves began, two dogs crossed the road, their ribs pressed against their mangy coats, mouths hanging open. Bill threw a stone, and quickly picked up another. Their heads lolled aimlessly but they did not come nearer. Bill leaped up the steps of the new police station two at a time.

Peeling a cucumber was a pleasantly idle thing to do. The narrow strips of dark green skin fell in the dust at his feet. Christopher offered a chunk to Adam who smiled but shook his head, so he ate the cucumber himself with a piece of feta, using the bread as a plate. He gazed down on the oasis of orange trees in the river valley, a paradise in this arid landscape. When the Minoans chose to be buried here, was the country different? Were there trees, wild goats and running rivers? Or was that how they wished it were, imagining on their seals a better world? Christopher had already walked the hilly summit looking carefully for ant holes, a trick he'd picked up years ago from a workman in Cyprus. The hollow chamber and *dromos* of a tomb were a dream to ants, offering them a startling amount of room. Thank goodness he'd shaken off the flu quickly. Bill needed him. Although he still felt a little weak and light-headed he was all right. He'd enjoyed the bus ride over. It was always amazing how different the south coast of Crete was from the north. It was as if the narrow isthmus which divided central Crete from east Crete was a frontier. They were now in Africa. Everything smelt different. It was the smell of rotted fruit dried to straw, the wind more parching than it ever was on the other side where there were olives and vines and purple hills. The only purple this side was the peak of Mount Dikte on the horizon.

Bill had been surprised to see him. He had not expected him on Sunday afternoon and Jenny rebuked him for coming too soon. He found her first, in the kitchen helping Margaret cut up tomatoes, her hands dripping when he leaned over to kiss her. Bill was in the pot shed, with most of the team sitting at trestle tables, stamping labels. The drawing boards were stacked against the wall ready for the morning. Bill's new supply of notebooks and graph paper was piled on a chair, with Bill about to dole them out when Christopher appeared.

Much later Sunday night Bill banged into Christopher's room and blurted out, 'Are you angry with me? I mean about supervising the tombs. Do you want to switch with me and carry on with your trench from last year? I don't mind. I was going to put Mary Elizabeth in charge of your trench and tomorrow put her to clearing around the threshold stone and starting a trial at the northeast corner. I'll be keeping an eye on her, of course, but I thought she would be the most reliable.' As so often when embarrassed, Bill's eyes scanned the ceiling, his hand blindly rummaging for a cigarette. 'Of the two of us you're the better archaeologist,' he rushed on, making less and less sense. 'Why haven't you ever had a dig of your own? Don't you want the responsibility? Or what?' By now Bill was blinking from tiredness. 'Have I ever asked you that before? I can't remember.'

Yes, Bill had, years ago on one of those evenings in the Pindus mountains when they first worked together on a dig, Bill even then set on having a dig of his own, convinced he'd run it well. Up there in the mountains he'd been terrific. He'd made them a fire in front of their tent every evening and brewed mugs of brandy and tea. He'd sung Negro spirituals in his good bass voice, and he'd described the life of a stockbroker which he had escaped, slapping his leg as he conjured up a fat capitalist with puffy eyes and a fat tummy, shut away from life in his carpeted office, sleepwalking through bear markets and bull markets and three-course meals.

'You should. You'd be good, people like you,' Bill pressed on.

'I don't like telling people what to do.'

'Don't you?' Bill sounded surprised. 'I don't mind. Someone has to do it. People can think of me what they like.'

Poor chap. When they'd found the yawning hole in the cliff the next morning, Bill had wilted. Christopher hoped Jenny had managed to calm Bill down before he went off to telephone Koraes. It was lucky Bill had such a loyal wife. Christopher couldn't forget how solicitous Jenny had been when he came down with flu. She could be so gentle when she wasn't worried. Bill was jolly lucky.

'I feel guilty. I don't want to make things easier for him. I want to make everything difficult, complicated, impossible . . .'

'Watch out!' Laura interrupted, frightened by steep drops where the road had fallen away inches from their wheels. Jenny bumped over rocks and lurched through gullies as recklessly as Laura's taxi from the airport.

'I stand in his way.' She pounded the steering wheel like a fractious child.

'But what's this got to do with the tomb?'

'He's so upset about it!'

'Poor chap,' Laura mumbled, guessing by Jenny's set expression that she wouldn't appreciate criticism. She certainly drove with dash, unafraid of the bad road. The canyon they were passing through, of low, isolated hills, their jagged summits hemmed in by higher mountains behind, was peculiar, even beautiful in the morning sun; it was still early and there was gold in the light, the sandy rocks not yet drained of all colour and shadow. Laura was not so excited by the landscape as she had been on Saturday and could converse with Jenny without regretting that their chat was losing her a moment of joy. Still, the country was astonishing. 'What do you think he should do?'

'It's so embarrassing!' Jenny's anger gave Laura a jolt.

'But it's not Bill's fault.'

'They discovered the tomb in April. It was a compliment to Bill that the Ephor left it for Bill to excavate. But Bill couldn't fly out from England just like that. The season was planned for now. I suppose he should have organised some kind of guard, but how could he from England? Now he's got to find out who robbed it, on top of everything else. The Ephor will be furious and who knows, Bill might lose his permit. They'll blame *him*, you know.' Jenny was near tears. 'And it's so unfair that it should happen to Bill! He's always said they trust him more than any of the other foreigners. They'll stop trusting him now.'

'Aren't you slightly over-dramatising the whole affair?'

'Tombs *are* dramatic affairs! This one was *un*robbed. If Bill had dug it in April, who knows what he might have found.'

'He couldn't.'

'He should've.'

'But how?'

'If you're to be trusted you have to pull your finger out. You can't sit back in Little Compton and have the world wait until you're ready.'

'You're blaming Bill. You're as bad as the Ephor chap.'

Jenny nodded. 'Fascinating man. Bill's let him down.'

'Nonsense!'

Jenny hit the wheel again. 'I know it's nonsense. That's why I feel so guilty. And why did Margaret suddenly refuse to come with me?'

'Miffed.'

'What about?'

'She felt left out. None of us spoke to her when she walked through with the frozen chickens.'

Jenny had to slow down for a three-wheeler putt-putt loaded down with sheep. Reeds on the right were making their icy sound. Tractors, donkeys, cars and more three-wheeler putt-putts had appeared, turning out of tracks to converge on Ierapetra.

'*I* feel guilty about the woman this morning who tried to sell me grapes,' Laura blurted out.

'That's not your husband!'

Laura laughed. 'No. But the ancient slippers on those old feet that dragged the donkey away when I'd pushed her out of the room make me feel ashamed. I was having my coffee. There was a screech and a hand came through the door swinging a bunch of grapes. She even produced scales and started to weigh the grapes in front of my nose. She couldn't understand that I had no money.'

'You get used to it.'

'I'm coming in with you now to change a traveller's cheque so that next time I can buy something.'

Jenny was taken aback. She caught Laura's eye. Laura smiled and squeezed her arm. 'I think it's awful about the tomb. Really bad luck.'

Jenny was mollified. 'Bill must be worried that one of our own workmen from Knossos is part of a ring. It was a professional job.'

'Does Bill know that?'

'He said so.'

The car had come to a standstill behind a donkey. Jenny frowned at Laura. Both women burst out laughing. 'Men are ridiculous,' Jenny swore, gripping the wheel with new energy. 'I sometimes . . .'

'You sometimes what, darling?'

'Hate Bill.'

Laura had to run to keep up, her flimsy sandals not as efficient as Jenny's ugly but robust ones. They'd reached the Demotic Market where Jenny said they'd begin, leaving bank, chairs and *stamnes* until last. Apparently they were making for the butcher, but Jenny stopped at the first vegetable stall when she saw oversized radishes that looked like red and white parsnips. 'I'd better get out Margaret's list,' she snapped.

Laura stood by, out of breath. Jenny handed her her passport, wallet, account book, biro and Kleenex as she rummaged for the list. Margaret's crumpled piece of yellow paper was extracted from the bottom of the bag.

It was a terrible list. It was both sides, and Margaret's handwriting was a spiky gothic which Jenny had trouble reading. She seemed to want all sorts of herbs and spices and pine nuts and lentils and chickpeas, as well as everything else. Jenny was incensed. Laura told her to calm down. 'I don't even know what a pine nut looks like!'

Laura suggested they visit the butcher. Meat was the easiest.

'And the most fun. Wait until you meet my butcher.' Batting the air with Margaret's list, Jenny led the way.

'*Kalos orisate!*' shouted Jenny's butcher, extending a big hand across the slice of a tree where he chopped the meat. Jenny blushed and shook hands, less loudly responding with '*Kalos sas vrikome.*' Laura extended her hand as well. The butcher sent a boy off to the café for drinks. Chairs were produced. Jenny and Laura found themselves sitting in the middle of the crowd as the butcher's guests. Margaret wanted cuts and parts for which Jenny didn't know the Greek words. She and Laura rubbed their tummies to convey the idea of pork's belly. A leg of lamb was easy; the butcher slapped his thigh and shouted, '*Bouti!*' A fantasy ensued about the beautiful legs of young women and the ugly legs of old women. The butcher reached round his chopping block to compliment Laura on her legs. He tilted his head, tapping her shoulder. Laura laughed her deep fruity laugh which shocked Jenny when she noticed its effect on the butcher.

When they moved on at last, Jenny and Laura were weighed down with three of the blue plastic bags whose brash colour the Cretans carried everywhere. The lemonade had gone to Laura's head. She giggled, drawing in breath. 'Quite fancy *him*.' Jenny was frowning at the list, then took up her two blue bags and walked on. Laura braced herself, annoyed that Jenny couldn't enjoy the fun.

Jenny was in torment. Laura began to pity her. Jenny's eyes studied the prices, remembering carrots were cheaper at the stall back there, not sure though if they had been as fresh. She asked Laura if she could remember and Laura suggested she buy the ones in front of her.

'Can't you remember?' Jenny criticised her.

Laura moved away to watch an old man in the archaic costume of the black Moslem 'skirt' cutting into a pomegranate; two pale hippies in ragged shirts bent over the old man, watching his hooked knife biting into the shell. The old man gave each of them half. The red seeds inside looked poisonous. If only they were, wished Laura, acknowledging her dislike of these wandering, parasitic north Europeans.

'You carry on and I'll take these to the car,' Laura offered, keeping her distance from Jenny as agreeably as she could.

The shrill of cicadas stung the still heat, and even the patch of shade under the tamarisk tree where Laura and Jenny sipped retsina was not cool. They'd been a long time. The sun flattened the pavement to a deadening glare. But Laura wanted a second bottle and eventually the boy brought it. No one was about. The boy resented the middle-aged women ordering more. He slammed the bottle down on the table.

Jenny stared at the sea, her eye on a small fishing boat far out, a red shape mysteriously still.

'His archaeology annoys me. He's so oblivious of everything else. I don't see the point. He thinks he's learning about the Minoans. But is he? What can you know from a few building blocks and broken plates? If I were a Minoan I'd be jolly offended.'

Laura laughed. Jenny gulped wine and choked. Laura hit her on the back, offering her a Kleenex. 'Would you like it if someone pieced together broken milk bottles outside your back door and then pronounced himself an authority on *you*, said how conventional you were, how many husbands you'd had, that you were a meat-eating type, that green was your favourite colour, that you had enemies, that you were greedy? What would you do if you were married to someone who thought that way?'

'But you love Bill.'

'I can't out here. I should. If I don't I'll lose him.'

'Could you ever feel at home here?'

'Never. Perhaps if I were still in love with Bill. But I'm not. And that's that.'

Laura folded her hands in her lap, watching a cat under the table to hide her surprise and embarrassment, reminded of a similar conversation she had had once with her father about her mother. It was the Christmas vacation of her second year at Oxford when her father suggested they meet at Bertorelli's for lunch. They'd drunk two bottles of Pouilly Fuissé with their roast duck; Laura remembered her father in a red tie with black polka dots and his brown herringbone suit. He was a good-looking man with her high cheekbones and dark, wide-apart eyes. She remembered admitting that if he weren't her father she could fall in love with him. 'I don't love her any more,' he had explained, biting into his duck. Drops of dark orange gravy stuck to his chin. He was off to Ireland; Laura was to give a letter to her mother with his reasons, which were place. He hated their house in Farnham. He was not that kind of middle-class person who cared if windows were draught-proof. Her mother's double-glazing was the last straw. His home was his family's ruin on the Dingle peninsula where he would go that night and live the rest of his life alone unless he met someone who liked draughts and leaking roofs and had at least the spirit of an aristocrat. And when Laura rushed to defend her poor mother, accusing him of self-indulgence and cruelty, he laughed. He promised Laura that she was just like him. She argued. How could she be an aristocrat when she'd never had any money, always been on scholarship, never had another home but Farnham? 'Farnham is no place for love,' were her father's last words before he walked off up Charlotte Street, leaving his daughter to go home to Farnham by herself.

Her mother's blotched face had been his doing. Jenny frightened Laura. Her mother had sat in the same chair all the holiday and grieved. Memories turned bitter. The wall-to-wall carpet silted up with dirt. It chilled Laura now to see the evil in Jenny's casting-off of another. Laura had only seen her father once since, when she and Francis were married. He'd come from Ireland with his second wife, wearing the same brown herringbone suit.

'I want to finish once and for all any chance there ever was of being happy ever again.'

'Leave Bill?'

'I'll make him leave me.'

CHAPTER 7

Laura had forgotten she was to have a roommate although Jenny had warned her. When she clambered up the rubble heap to her room for a quick lie-down, there was Annabel, the antique dealer from Pimlico, grappling with the other camp bed in a straw hat with ribbons tied under her chin. She waved a hand laden with big rings and said, 'Hi.'

Annabel was the only person who spoke at lunch, everyone else was too tired from the first day of digging. She sat beside Christopher who ate his food with a wooden expression on his handsome face, which delighted Laura.

But Annabel was unaware that her voice intruded on everyone's fatigue. Perhaps people sitting round a table in silence compelled her to talk. Laura might have sympathised if she herself weren't so exhausted. Annabel had changed her clothes. She was now wearing an outfit Gertrude Bell might have worn if she'd shopped in the King's Road. She'd tied a black scarf around her neck, the ends draped over a high-waisted khaki riding jacket. A long blue gingham skirt reached half way down her calves, leaving her not very good ankles and black canvas shoes for all to see. A deep, butch voice came out of this buxom, dark-haired woman, a voice, Laura decided, with the hide of an elephant.

Jenny and Margaret set down plates of food. There was chicken fricassée in a lemony sauce, courgettes and carrots sautéed, a cucumber and tomato salad with an exotic sort of parsley decorating it – the 'coriander' on the shopping list. When everyone had been served, Margaret appeared with another dish, leeks in a cheese sauce. Jack groaned. Jenny looked furious. Laura held up her plate although she agreed with Jack.

'And Charles told me how months later this Italian chap turned up in his office at the British Museum with a package, that same marble head that he'd noticed at the back of the shop in Rome!' Annabel guffawed, dropping her celebrity's name with silly excitement. 'Of course Charles could recognise the fake from the shape of the package and told the man not to bother unwrapping it. The surprise on the man's face, Charles said, was delightful. I come up against such people quite often myself, actually. Really it's fascinating how forgers believe you'll buy their fakes. Charles and I have had some good laughs about it. Stolen property is a different matter, of course. I insist on a pedigree for the thing and check it out when I think the person is suspicious.' Another guffaw again interrupted the monologue. 'There was a chap last winter, a most plausible sort in a brown suit and thick brogues, who wanted me to buy his ormolu clock. It was a very beautiful one. You would have adored it.' Her red finger nails tapped Christopher on the hand. 'There was no doubt that it was genuine.' There was a bland expression on Christopher's face as he took seconds of the leek-and-cheese dish.

Then Laura noticed Jenny's clean plate two places up from Annabel. Jenny was not listening. Her green eyes were wide open and vacant. When Bill pinged his glass Jenny had already left the table.

'I should like to see site supervisors down at the *apotheke* at six. 'And,' Bill added, 'I've just heard that the Epimelete of west Crete will be visiting us tomorrow. So, everyone be ready for that.'

Jenny had even scrubbed the floor, and damp patches were still showing when Bill made it back to their room after talking with Christopher. She'd rolled up their sleeping bags so that they looked like eiderdowns at the end of their camp beds. On a chair by his bed she'd set his leather stud box, his brushes and his two World's Classics of *War and Peace* and *Ayala's Angel*. Propped up on orange-crates against the wall were their suitcases. She'd hung lamps from nails over the beds and unpacked his dressing gown and slippers which he hadn't been able to find the night before when he'd needed to go to the loo. Bill wondered what had come over her.

Bill called Jenny's name. He went out and knocked on the loo door but Annabel answered.

Bill went back into their room and shut the door. In the gloom he pulled off his shoes and lay down. It was a fact that Jenny knew a great deal about the Bronze Age, when she was in the mood. He'd overhear her at dinner parties tell someone about the Minoan palaces and Minoan pottery and Egyptian imports – she could even describe a faience jug. But the past revealed stone by stone on top of a windy hill was not her cup of tea. All the same Jenny did not think two thousand years ago was irrelevant. She was as glad as he was that he wasn't a stockbroker. If she could only change her mind about excavating.

It was Badger, an uneducated peasant, who had converted Bill on his first dig after coming down from Oxford. Bill decided quickly that Badger was a better archaeologist than the director, his father's partner's son, who would rush from trench to trench in a frenzy. Badger taught Bill to sense with light taps of the pick how the earth was changing. Down Badger would go, layer after layer, telling Bill what to write in his notebook, when to draw his section, showing him when the ground was hard and they'd reached a new floor. He'd hold out a sherd, pointing out the decoration, and date the floor for Bill. Badger had learned all this without going to university, but unlike his father, Badger was not in awe of scholars. Which freed Bill from his father's fantasy that scholars were wise and wonderful people. Badger had worked for the English for years and knew how strange and wrong they could be. He could even remember Sir Arthur Evans. But he'd say it wasn't the English nor Sir Arthur but the Cretan earth that had taught him what he knew.

Bill kicked the eiderdown off the bed and punched up his pillow. He loved Jenny. He'd never loved anyone else even if he sometimes flirted with a pretty supervisor. He was always discreet. Last year he'd only kissed Claire a few times, walking late back to the site to check on the tent and discuss her trench. Jenny was wrong to think they'd gone further. He had found Claire pleasant to be with. It was lovely to feel her young shoulders and smooth ribcage against him.

Why wasn't Jenny having a nap? It soon would be too late.

Bill had dozed off when the door juddered open. Jenny dragged her feet across the floor and kicked off her sandals. The camp bed went 'zonk' as she flopped down onto her back without a word, her eyes on the ceiling. Bill raised himself on to an elbow to make her look at him. But she wouldn't.

'Thank you very much for doing the room.'

Jenny turned her head to look at him.

'Darling, really you were terrific to do it because you must be exhausted.' Bill stretched out a hand.

Jenny sighed.

'How are you? I talked to Koraes who was actually quite nice. Said a few curt "yes, yes, yes"es when I told him but didn't blame me. Which was a huge relief.' Bill lay back down, tucking both hands under his pillow. 'He might even have been expecting it to happen. From the way he sounded not too surprised I wonder if he didn't already know.'

'Did you now?'

Bill was back up on his elbow. 'You aren't angry with me, are you? I was going to do the room. I was prepared to when I found you'd already done it. Come out to dinner. Tonight. Why not?'

'Margaret's made pork stew.'

'Then after I've spoken to everyone let's take a walk.'

'I'm too tired.'

'By the way, I was wondering just now if you'd like to supervise your own trench?'

'Do what?' Jenny sat up abruptly and pulled her pillow around so that it lay in her lap.

'You might enjoy it.'

'I've never supervised a trench in my life!'

'It would get you out of the kitchen.'

'It's boring.'

'What is?'

‘Supervising a trench. It couldn’t be more boring. I’d hate it.’ She was pummelling the pillow.’

Bill lay down again and turned onto his side. He shut his eyes. He needed sleep. A little nap, before he had to explain to the supervisors why labels were so important. Why Jenny had done the room was one of those things. He yawned, just one of those things, grateful he’d not had to do the room himself. Jenny had had no change of heart. Forget about Jenny until there was more time and less to do. Put her right out of his mind. He couldn’t afford not to.

‘Darling Christopher.’ Laura had asked Annabel after lunch if she was Christopher Bendick’s friend. Laura was thinking Christopher must have been cornered by this freak at a drinks party in London. She’d have gone on at him about her dream to work on an excavation. She’d noisily unpacked all the afternoon and pushed her bed about although Laura was trying to sleep. But Laura didn’t dare tell her to shut up. It might be twisted into an outrageous insult Laura had hurled at her when she was unpacking as best she could in the murky bedroom. Laura was wary of such women.

Had she decided that archaeology was the way to Christopher’s heart and come out here to dazzle him with her enthusiasm? Could he be dazzled? Certainly Laura had taken in, as Annabel was sure to have done, his wide shoulders and narrow hips.

At supper the only chair unoccupied when Christopher arrived was on Laura’s right. Christopher tilted the chair back and threw his leg over it. Annabel pretended not to notice. She bombarded the boy Edward with questions. Hiding her triumph, Laura only tilted her head in Christopher’s direction; her eyes she kept riveted to the mound of pork stew in front of her. She asked him without shifting her gaze (except for a quick glance across at Annabel) how he was feeling.

Christopher was delighted at his luck. He liked Laura’s impulsive gestures. She’d been a fool when she fell off the lorry, but it was the foolishness of a clown. That kind of courage moved him.

Christopher filled their glasses and started in on the portion of stew which Jenny had handed him immediately he was seated. Out of the corner of his eye he was relishing again, as he had the day they loaded the lorry, Laura’s unruly hair, a defiant curl brushing his ear, the points of her high cheekbones, and her wide, turned-up nose bent over the food like a hungry child. Was she wild? His mother hated hair that looked uncombed.

'I hope you won't mind helping me with the tombs,' he began. Laura brushed back the hair with her hand. She might as well not have bothered.

'What did the Ephor say? Bill's keeping it all such a secret!'

'Nice man, Koraes. The Epimelete of west Crete is coming tomorrow.' Christopher's jaw worked impassively on a chunk of meat. His tone was as expressionless as his eating.

'But Bill was so worried.'

Christopher smiled.

Laura retreated. 'Are you really quite over your flu?' She changed the subject.

Christopher tucked into more stew. 'I'm fine.'

It was Laura's turn to smile at Christopher's modesty in his expensive golf shirt, another in what must be quite a collection. The yellow wasn't as becoming as the green, but the wide sleeve hung well. Was he vain? His good looks were unfortunately of the sort Laura had never found attractive, the straight nose and regular mouth too obvious for her. 'I'd love to help with the tombs, if I can do it properly. Whatever you and Bill think . . . You have the other room by the sea?'

Laura felt Adam's eyes on her. He was leaning back in the chair opposite with his arms crossed. He'd refused food. He was smiling. Laura wondered if he was falling for her. It often happened. She attracted men, unlike the luckless Annabel who could only attract a frightened Christopher type, although even that seemed improbable. Laura caught Adam's eye. The dull light from the storm lamps was restful and protective. Laura smiled back, encouraging him.

'Every summer when I was a child we went to Cornwall to stay with my mother's sister who had a marvellous house at the top of a cliff. The waves crashed against the cliff but it was so far below my bedroom window that it sounded faint and safe, and exciting. Do you know?' Christopher prattled on.

'Where were you born?'

'London. Very ordinary. I always wish I'd been born in Cornwall.'

Laura laughed; Adam in the shadows opposite swallowed wine. 'Do you wish many things were different about yourself?'

'Don't you?'

They both laughed. Laura pushed her plate away and moved her chair back. 'Practically everything.' She was laughing so much, tears were coming. She'd drunk too much. She reached for the jug to refill her glass and offered to refill Christopher's. He held his glass to the lip of the jug.

'I make mess after mess of my life,' she burbled, leaning nearer.

Annabel had just asked Edward what he'd read at university. She described her friends who'd started out as solicitors. One was something in the National Gallery, another was the linch-pin of the Arts Council, and several others, though bankers, were influential in other spheres. 'Make sure you pass those exams,' she harangued Edward, admitting that even her father started out that way. She wore a black velvet ribbon around her neck, a red velour jersey with plunging V-neck and puffed-up shoulders, and something that glittered in her hair. Bill, further down, was dreading a change in the weather.

'What sort of messes?' enquired Christopher.

Affronted by the urgency of Christopher's question, Laura fumbled with her knife and fork. 'Let's switch to tombs. You tell me what I should expect of the Minoan way of death.'

'Gold!' Adam shouted. Laura and Christopher squinted into the shadows.

'What was that?' asked Laura.

'Ask Christopher. I'm off.' Adam heaved back in his chair. 'I don't eat cheese,' he sulked as he bumped into Jenny, who whispered to Bill and followed Adam out of the room. Laura shook her head, watching their exits. Christopher heard her sigh. He agreed. In fact he wished to talk with Laura some time about Jenny and Bill. Adam mattered less though he'd surprised Christopher by how efficiently he'd planned the new trenches.

Jenny came back through the door alone, colliding with Jack and Susan who were just leaving. She screamed after them that a rota for washing up was pinned to the wall; Jack escaped with Susan right behind, both holding their ears. Ellen nearest the list read out Edward's and Mary Elizabeth's names. The scrape of chairs and shuffle of feet in the crowded room emptied the table quickly of cutlery, plates, glasses and stray pieces of pie crust. Mary Elizabeth's drawl was heard through the partition wall as she marvelled at Margaret's 'best cooking'.

Laura and Christopher sat on. Christopher refilled their glasses. They both stared at the bare table. 'It's like on an aeroplane when everyone's eager to get off,' Laura murmured so that only Christopher could hear. Christopher nodded. Jenny interrupted them, needing to wash up the wine jug. Christopher handed it up to her, avoiding her eye. Laura got to her feet, suddenly eager to be through the door before Christopher could catch up.

CHAPTER 8

The Epimelete showed no interest in the tomb. Bill was so relieved that he didn't visit the cemetery hill for two days, leaving it to Christopher. Christopher shovelled out more and more of the loose earth from the *dromos*. He found a ten-drachma piece which had fallen out of one of the robbers' pockets, but no whole pots. The robbers had done a thorough job. Laura hurried from Christopher to Susan, backwards and forwards across the cliff with bags and labels, the knife or the smaller trowel. It was mindless work. By three o'clock Margaret's huge lunches were just the thing and the exercise meant that she could eat as much as she liked. She still fitted into her size twelve jeans. Nor had she ever before experienced the long, sound naps which ended her afternoons.

It was Laura who spotted Bill's red car at noon on Thursday, inching forward, obviously fearful of the stones and ditches and loose earth – a different sort of driver from Jenny. Laura waved and Bill's hand, looking very small, waved back. Bill stuck his head out of the car window to shout something but when it reached Laura it was too faint to make out. She turned back with her metre tapes to wait for the next thing Christopher wanted to plan. It was another beautiful day, dusty and hot but clear. She shut her eyes, tilting her face to the sun. Bill would not reach them for some time. When he did there wouldn't be much for him to see.

Laura felt the heat sink into her. Back in London, John would be running in the rain right now, his collar up, diving suddenly through a doorway for a drink and a sandwich, with wet feet. He was so nimble and abrupt in the way he turned a corner, entered a building, crossed a street. She loved when he moved, how he was in command. If they had just said goodbye he skimmed past people like a bird. What a pity he loved his wife and two children too much to start out on something new with her. Wouldn't it be more fun to be out here with her right now? Much more fun for her to have him loving her here on this exotic island with all these archaeologists. John's face could still bewitch her as she pictured it now with her eyes shut to the sun – squashed up like a monkey's, his long flat nose nearly touching his thick lips. She hoped she could forget it . . . Always dressed impeccably in dark suits and cream shirts, that face promised what the clothes covered.

'Great news!'

Laura swung round; she'd forgotten Bill. Her elbow knocked him in the chest. Why did he creep up on them? She felt giddy from the shock, and hunted for somewhere to sit. She cursed Bill, who smiled.

'Sorry.'

'You're not at all sorry.' Laura glared at him.

Slowly Christopher climbed out of the *dromos* and brushed off his trousers.

'Christopher, guess what we've just found in your trench?'

Christopher shook his head, a stern stare putting Bill on guard.

'A staircase, ashlar blocks, two so far. With edges crisp as crisp! Looks new. You were right to call the courtyard palatial. I admit it.'

Jenny had insisted on a walk and then marched off in a mood, leaving Laura to finish the coast road alone. When Laura was back in the village the faint light from candles and paraffin lamps lit up the doorways; inside men sat at tables drinking. Outside in the warm darkness others sat smoking, odd chairs and tables left about in the street. Slowly Laura walked on, uneasy as she passed the cafés in case she'd be seen.

What was this madness that turned all of them into caricatures of themselves? It bewildered Laura. Was it the suspicious, inquisitive eyes on them all the time that compelled the English to forget their dignity? Was being foolish a way of protecting themselves from these strange people and each other? Jenny's misery was so undignified, it made it hard to remain loyal.

And Christopher surprised her. She would not have believed he could be so absurd. It had been after Bill left. She asked what they might have found in the *dromos* if it hadn't been robbed. 'I wasn't the robber!' he snapped. All week she'd let Christopher order her about. She supposed it was Bill's visit that made him be so rude. But he did look annoyed about the staircase. And the work of cleaning up after the robbers wore him down. After Bill left Christopher had been silent, until he blew up at *her*! She had walked away. She'd been trying her hardest to please him, and it hurt to be slapped down like a dumb fool.

Then, from the shade of the olive tree where she went to recover, she heard shouting. Suddenly earth spewed into the air, fresh bursts erupting from a shovel in the *dromos*. Laura went back, too astonished to keep away. She found Christopher kneeling at the edge. He was quite changed. No longer were the corners of his mouth right down with a deep frown creasing his perfect looks. He'd turned into a rowdy schoolboy, slapping his leg and brandishing his pick to urge on old Andonis, the strong and stupid shoveller who was no archaeologist but Christopher's great friend. Andonis shouted at Manolis to hurry up with the wheelbarrow and Christopher shouted also like an adoring copy-cat. Laura sat down on a flat stone and watched Christopher let the old man make a fool of him. Her unease embarrassed her. She kept glancing up at the empty sky for relief.

At lunch Christopher had been in terrific spirits, telling an outlandish story about his grandmother who had been expert at betting and who on her deathbed had tipped him the winner of the Derby. Right away he'd put on a hundred pounds and come out of it a thousand ahead, which more than paid for his grandmother's funeral. He let out such a guffaw that everyone looked. It interrupted Bill's diatribe on palatial staircases. Ellen and Margaret frowned at their salad but Jenny laughed.

'Nice walk?' A pale T-shirt accosted Laura in the middle of the street. She couldn't see the face but she recognised the young voice. The narrow form wasn't Christopher's, nor was the pale T-shirt. She registered her disappointment and tried to be amused at herself wishing it weren't Adam. Adam offered Laura a drink which Laura had to accept.

Adam had decided that Laura was the only person in the group who could be his ally. He liked the way the lines at the corners of her eyes and down her cheeks gave her face mystery. He'd gone off young smooth faces which were shallow and careless of other people. Not Laura.

They found a table near the door of the second café. There were fewer people here. From inside they could hear the café owner's wife's rasping voice reverberating from the bare walls. Eventually the woman brought out the ouzo bottle and two glasses.

Adam knocked his glass on the table in the Cretan style and drank to Laura's good health. Over the rim of his glass he spotted Christopher two tables away with the village schoolmaster. Had Laura seen him? He hoped not. He'd been watching Laura and Christopher at meals. Laura often laughed at Christopher's remarks, or came back with some long remark of her own. Fortunately her back was to him.

'What do you think of the village?' he asked, fixing her with his eyes.

'It takes getting used to, I suppose.'

'It's filthy!' Adam pounded the table, which Laura hadn't expected. He went on, 'The fear is horrible. Don't you feel it? And those old women who are the widows of the men the Germans shot. They stumble back to their hovels without anyone to help them!'

Laura listened and was looking at him now. 'Why do you say there's fear?' she asked.

Adam held up a clenched fist. 'It's a dictatorship. Sons betray their fathers. You see whispering. No one's safe.'

'But aren't you exaggerating?'

'No. I keep my eyes open. I sit at these cafés and watch. Have you noticed the German tourist? He arrived yesterday. The villagers hate him. They'll kill him if he doesn't watch out.'

'That's silly.' Laura looked about her now.

'Why don't you believe me?' Adam pleaded.

'Look, there's Christopher.' She waved. Christopher called to them to join him.

'You haven't answered my question,' Adam reproached her. Laura stood up.

'Let's go and meet Christopher's friend.'

Adam fumed. He was slow to follow. When Christopher introduced them the village schoolmaster, a white-haired man with a stick and a solemn face, thrust out a hand across the table which Laura shook. Adam kept his hands in his pockets. The schoolmaster waved his stick at the café owner's wife who refilled their glasses, her grimy hands pouring liberal amounts from the two large bottles of raki and ouzo.

The schoolmaster stared ahead of him. Laura asked Christopher what their talk had been about. Christopher, in a low voice, avoiding the implacable look of their host, explained that the schoolmaster's wife had watched the Germans shoot her father, her uncle, and her two brothers. Adam tapped the ground with his foot. He knew the story.

'What does he think of Germans now? Can his wife forgive them?' Laura asked. Christopher translated. The schoolmaster narrowed his eyes as if he were about to take aim, and shot out 'never' in Greek, his hand resting as still as stone on the walking stick.

'*Pote*?' repeated Laura.

'*Pote*!'

Christopher mumbled something into his raki, looking stupid. Laura was angry. She kept brushing hair out of her face.

The schoolmaster started again. Low, urgent Greek poured out of him, directed at the listening smile on Christopher's face. It was only from the way Christopher gulped his raki that Adam guessed he wasn't comfortable. Did he think the schoolmaster ridiculous? The arrogant bastard. Adam squinted as he pictured the butchery. He sympathised with the schoolmaster.

'What is he saying now?' Laura tapped Adam on the shoulder. 'Can you understand?'

'He can't understand how Germans can come back here.'

'So he can pay them back,' scoffed Laura.

Suddenly an arm pushed itself between Adam and Laura. Something was clutched in the hand. Behind them a man laughed. The scrape of a chair being dragged and put down followed, and two big legs in pale trousers wedged themselves between Adam's and Laura's seats, the hand opening and dropping on to the table a little brown frog. It jumped against Laura's glass. When the German laughed into Laura's ear in broken English, 'Don't you think good joke? It is a little, little frog, you see,' Laura reached out to cup her hands around it and lifted it from the table. Then another strange pair of hands appeared at Laura's other side, and a light, American voice said, 'Give it here.' This man had also squeezed into their group, pushing thin, blue-jeaned legs between Laura and Christopher.

Adam was appalled but admired Laura for rescuing the frog. Christopher impassively drank and looked nowhere, his eyes fixed to a spot not far from the schoolmaster's nearly full glass of beer. Nor did Adam dare meet the schoolmaster face on, stopping at his papery hand gripping the knob of the stick.

'Drinks for everyone,' shouted the big German, waving at the café owner's wife. Laura handed the frog to the American, who buttoned it inside his shirt pocket. The café owner's wife ran away up the two steps into the bright room of the café. 'You are my friends,' exclaimed the German, waving his big arm in the dim light of the street. 'I like you!'

Laura lifted her chair back so that she could get out. She hurried inside after the woman, returning with the two bottles and glasses. The schoolmaster's eyes followed her as she set a glass before the German and offered him a choice between the two bottles. She smiled at the schoolmaster, whose full glass of beer sat in front of him untasted. Then carefully she tilted the ouzo bottle towards Adam's empty glass. Adam pulled the glass away so that the ouzo poured onto the table.

Abruptly the schoolmaster stood up and walked off. Adam stumbled over to the schoolmaster's place, picked up the glass of beer and threw it in the German's face.

CHAPTER 9

When it turned out that, after lunch on Saturday, only Christopher and Laura had decided not to go anywhere, Christopher touched Laura on the arm just as she was rushing to her room for a nap and suggested that she come down to the pot shed where his room was and have a drink with him later. There was much to talk about, he reassured her, when she smirked. It would be business. So Laura decided not to change into something less drab, only feeling cleaner in her better jeans and new, yellow T-shirt as she made her way to the pot shed. Probably it was Adam and the German that Christopher wanted to discuss, Laura thought, preparing herself for a serious talk. It wouldn't surprise her if Christopher was afraid the German might pay Adam back for his outrageous behaviour. Should Adam somehow be persuaded to apologise? Would he?

Christopher was bent over Jack's drawing board when Laura arrived. He asked if Laura could possibly find another glass in his room while he quickly finished what he was doing. She went on through to look for the glass. Immediately she spotted a glass on the floor by his camp bed, and then, stacked beside it and on his chair, was paperback after paperback of dirty novels. She glanced behind her through the door at the figure frowning over Jack's drawing. Could it be? It couldn't be; certainly this was not the reading of a superior intellect! Tresses of blonde hair Veronica Lake style clothed salacious nudes on the covers; lips, bosoms, a shadow of pubic hair were the inviting background to the boldly embossed titles of *Bangkok Girls* and *Reeperbahn Overnight*. She paused to riffle through them, delighted to find such riotous covers in Christopher's bare room. She'd never have guessed. And very funny. His jacket and trousers hung neatly on nails, the rest of his clothes tidied away out of sight in the large blue Globetrotter suitcase at the bottom of the bed. Only the paperbacks were in a tumble.

Laura stooped down for the glass and tripped back into the adjoining drawing office. From the bland way Christopher thanked her and poured retsina into the glass, Laura supposed he'd forgotten those books were lying about. Or did he not realise the import of such literature?

Brimming with delight, Laura wanted to find something else to laugh about that couldn't offend him. A full circle she completed, twirling herself on the stool in her jeans, holding up her glass, calling 'cheers'. She had been amazed, walking down after a good nap, by the dreamlike whiteness of everything, the pale sea melting into a white sky, the rough road above the beach (where Jenny had been so unpleasant on Thursday) brilliantly white. As unreal and intoxicating as was this discovery about Christopher. Pearl-white clouds were shrouding the dark hill behind and hung low over the dazzling road, wrapping the place in secret security. And they had the place to themselves. Ellen, Adam and Annabel had gone to Zakro with Jenny and Bill. The others had left on the three-o'clock bus to Ierapetra.

'Well.' Laura giggled. Imagine Christopher lying stiffly in bed with his impeccable profile bent over steamy descriptions of melting breasts and burning tongues! If only she'd known, she'd have put on a dress. When she had thought Christopher was resenting Bill, or pitying himself with the German and the schoolmaster, was he actually imagining a good fuck? How wonderful. 'I wonder what that German's doing,' she said idly, willing herself to take control and not frighten Christopher with too much amusement too quickly. She gulped down the retsina and held out her empty glass. 'I feel like I've been let out of school. Wasn't Adam awful to throw that beer at him? What came over him?'

Christopher sat hunched on Jack's stool, his arms hanging straight and stiff between his legs. He shook his head.

'But what did he do it for? That poor German.'

Christopher looked up.

'I think the German was jolly brave to take it like he did. Adam deserved a good thrashing. Who was he trying to impress?' Laura tilted her head enough that she could peer at Christopher out of the corner of her eye. She adopted a conniving tone, delicious mischief changing the world for her at this moment as she sat in front of Adam's drawing board. 'What do you make of the boy? Is he just young? Or what?'

'He's smitten with Crete.' Christopher threw back his head. 'Or you.'

'What's funny?' Laura pretended to be incensed, delighted by Christopher's insinuation. 'I don't see anything funny. The German tried to make friends, even if he was going about it in a rather strange way. I didn't like his American friend. Calling us "damn Limeys" as they walked away wasn't very nice.' Laura started to giggle again, sniffling into the Kleenex she always kept in a sleeve of her T-shirt.

Christopher suddenly, preposterously turned serious. He leaned back on his elbow, the corners of his mouth right down. 'I wasn't enjoying it very much but the schoolmaster is important in the village and has been a great help to us. His collection of antiquities in the schoolroom is excellent.'

Not to be outdone, Laura stopped giggling and returned his serious look, breathing in to accentuate the seriousness of what Christopher had just said. She was discouraged all the same.

'But I don't like the schoolmaster. Of course his hatred of the German is horrible.' Christopher frowned. 'The German meant no offence and the schoolmaster's behaviour was extremely unpleasant. I couldn't agree with you more about that.'

Laura uncrossed her legs and smiled.

When Christopher climbed off his stool, dry mortar skittered down. He shuffled across the room uneasily. Laura had draped herself over her stool so that one of her long legs stretched straight to the ground, her thin thigh and bare ankle giving her precarious support. Christopher had never thought he liked jeans. There were too many of them about, but Laura wore hers so off-handedly – as if they were a quaint necessity rather than the modern look – that he was thinking he preferred her in blue jeans to anything else. Which compelled him to move from his stool.

'I wonder where we'll find food. I've never tried in the village before.' He changed the subject, peering closely at the blossom bowl which Jack had drawn and which was on the table still; he picked up the callipers and checked Jack's measurements of its carved lotus-leaf decoration. He would not listen for the creak of Laura's stool but bent his whole attention to the accuracy of the drawing.

'And how is it? All right?' Laura was beside him, her cheek touching his shoulder as she watched him correct the size of the leaves. Nervously Christopher bit his lower lip, then suddenly straightened up with a histrionic sigh – out of relief, he pretended, that another job was done. He dropped his pencil deliberately and looked for the bottle of retsina. If Laura hadn't stepped back he would have bumped into her. She stepped back.

'I'd love a walk. Would you?' Laura asked, following Christopher's dash for the wine. Men for Laura always started as a challenge; a very ordinary reaction to the opposite sex she knew, and she regretted its ordinariness, quick to notice when other women tested their attractiveness in this way. Annabel, for instance, was a clumsy, incorrigible flirt, and the same age as Laura. But it was fun and the comedy of a conspiracy fitted in well with the dream world outside. Now that the unattainable Christopher turned out to have the same thoughts as other men, Laura could not stop herself.

The unattainable side of Christopher was becoming more pronounced by the minute. How direct could she be with him, Laura now wondered. She found herself becoming more diffident faced with his fear. He was afraid, poor man. He'd left his stool because of her, not because he was hungry. She began to feel sorry for him. She quelled her mischievousness and pulled back her shoulders, holding out her glass.

'Fine. Why not?' Christopher agreed about the idea of a walk. 'It's too early to eat.' But then he sat down on Jack's stool and stared at his shoes. 'Need a polish,' he mumbled, and withdrew into a glum reverie. Laura went back to her stool.

Christopher wished Laura would not give up flirting. He stared at his shoes wishing she'd keep at it. He liked her. He had liked, on Thursday at the café, the way she went and fetched more drinks for everyone when the wife wouldn't serve the German. He agreed with her about the German, and Adam. He agreed passionately. Laura was so right to hate the schoolmaster's unforgivingness and Adam's foolishness.

If only Laura would encourage him. 'Please, Laura, encourage me,' he cried to himself, a silent man on a stool. Would she? His dusty shoes held his eyes only because otherwise he'd see Laura's frail wrist exposed by the glass in her hand. He longed to look up. Should he? He would. The sizzling whirr of the Camping Gaz reminded him that they were wasting the gas. He crossed to Laura's stool and unhooked the lamp from the nail. Laura's face was directly below him. He'd carefully avoided brushing her when he reached for the light. But her smile intoxicated him. 'Is Jack's drawing terrible? Is that what's wrong?' she asked.

Christopher bit his lip again and shook his head. Laura bit her lip and shook her head back. 'No?' she teased, still looking up. 'That's not it?' Laura kept still. Awkwardly, that expensive sizzle which he had meant to cut off swinging from his left hand, Christopher leaned down and kissed the intoxicating smile.

Christopher turned off the light and in the dark groped for the latch of the door; it creaked open, the old boards bumping against the rough floor. Earlier it had been like walking in moonlight; now it was pitch black outside. Laura asked if Christopher had a torch, afraid of what she might fall into or step on. The sky was as black as the ground. The leisurely fall of the waves sounded near; a rooster crowed, a donkey broke into asthmatic gasping, ee-awing interminably. Then a black stillness settled again over this end of the village which was as far from the cafés as it was possible to be, the pot shed the last building to straggle into the barren scrub.

To find the torch Christopher had had to light a storm lantern. At last he found the torch under the books in his bedroom while Laura waited in the doorway.

The circle of light flitted from stone to wall to doorway; twice Laura stumbled, unable to see what was in front of her because Christopher couldn't hold the torch still. She clutched his arm but did not complain, wary of the effect it might have. It wasn't that she was happy, but she was interested and concerned, alive again, thank goodness, to another person's quickest feeling which, like a hair crack in a polished world, dissolved the symmetry of Christopher's appearance minute by minute. She was excited, curious to see what was to happen next. His kiss had been so tentative that perhaps it had been a mistake. Certainly it was no beginning to anything. Not yet.

They were in search of food. Christopher seemed with his torch to be looking in strange places. They passed the side of a rusty oil drum, a heap of stones, a bent wire fence. Up and up they climbed until they reached the street with the cafés.

Christopher switched off his torch, marooning them in the dark middle of the street. 'None of them has food,' he moaned, unaccountably dooming their search at this point. Then suddenly the raucous guffaw of the German reverberated from the inside of the nearest and most brightly lit of the cafés. Christopher grabbed Laura's hand and hurried her along to Vasilis's. Laura was more and more astonished but not angry. When they reached Vasilis's she refused to go inside, settling herself at the table furthest from the door. Quite enough light, however, to drink by. Laura wished to stay with wine. Vasilis brought them a cut-up cucumber and salted almonds. Christopher, Laura could see, was picking through the nuts nervously, eating one after another. Laura laughed and slowly bit into a piece of cucumber.

'Hmmm. Good.'

'What?'

'This cucumber's delicious. Don't need anything else. A rest from Margaret's cooking will do wonders for the figure.'

'Oh.' Christopher sounded disappointed. Laura held up the plate to him. 'You try,' she urged.

'I was hoping for more to eat than just this.'

'Tell me, Christopher, about your life.' Laura poured out more wine. 'How do you spend it exactly? I mean do you work or anything?'

'I work. I'm not paid for it but I work. I like the opera and go when I can. Both my parents are dead, so I haven't any of those sorts of responsibilities.' Christopher offered Laura the last nut which she took, her first. She wondered if Christopher could ask Vasilis for more. Christopher was relaxing. With alacrity he walked to the lighted doorway with the empty plate and called, laughing pleasantly when Vasilis cracked a joke.

'What was that about?' Laura asked when he sat down again.

Christopher chuckled and shook his head. Laura imagined he was blushing. What an extraordinarily shy man he was. She'd often had success with shy people, doing most of the talking in the beginning. Or asking questions. But not too many. Too many questions frightened shy people. It forced them to be articulate about themselves and that was not a good thing since they'd fail and be exposed floundering in public. If she made a fool of herself, that helped. The shy could laugh then and momentarily forget themselves. Was it an intense self-consciousness that was at the core of such fear? She thought not. Fear of some sort. Fear of rebuff perhaps, and it wasn't self-conscious to be afraid others wouldn't like you. Laura knew how very eager she was that people like her. She hoped Christopher didn't regret that he'd kissed her – if in fact he'd meant to kiss her. But something had happened. He'd become abrupt and withdrawn and pessimistic. Was it because they hadn't found food? 'Are you starving?' she asked, offering him the last piece of cucumber.

Christopher was eating through the nuts again. 'You're eating those nuts as if you haven't seen food for months. You poor thing. Let's ask Vasilis. Couldn't his wife make omelettes? Go and ask him.'

The wary look Christopher gave Laura before getting up from his chair again made Laura burst into giggles. He'd been a bachelor a long time, obviously. He wasn't used to women's suggestions, but was polite and willing so far. When would he first refuse her something, she wondered. Already he'd been rude but she didn't think he'd meant it. Could he ever *mean* to be rude?

When Christopher came back to his seat Laura stretched out her hand, pushing it along the table and resting it palm upward in front of his place. Christopher looked at it. 'That was a good idea of yours about the omelettes,' he conceded, patting Laura's hand but not taking it. 'Maria's making them now.'

Laura withdrew her hand. 'Well done,' she retorted, the wooden top of their table a wide, rough space between them now. Staring at this uneven surface scattered with bits of nut and dark strips of cucumber peel Laura succumbed to Christopher's bachelor sterility; her hope of enjoyment shrivelled. He'd rejected her hand. Furiously to herself she told him to go back to his dirty books and wallow in vicarious thrills, but never again to mislead her with lustful looks and a kiss. What a pathetic man, she repeated as a comfort. Pathetic.

'So.' Christopher held up the empty beer bottle which Vasilis used as a carafe. 'Vasilis had better bring us more.' Laura thought Christopher's yellow golf shirt served him right, resenting its expensive cut. Brushing off crumbs, she patted her new, cheap, yellow T-shirt. It was *her*, like her jeans and her sandals. Impatiently Laura turned her chair so that she could stretch her legs out beyond the thick leg of the table. She reached up to catch strands of her hair and fluff them up. 'I think it was at this table that Adam gave me a beer when I arrived. Jolly nice of him really. I was taken quite by surprise. Apparently his mother's a friend of Jenny's. Did you know that?' Laura raised her eyebrows at Christopher. 'Strange that. Adam's not at all what one would expect from a friend of Jenny's. All that fresh Cotswold air you'd think would make for a healthier-looking person.'

'My father was a farmer.'

'That's why you have such a brown face. That's the first thing I noticed about you.'

'The first thing I noticed about you were your legs.'

Laura had tilted her chair back. With a jolt she righted herself and kicked up a leg, chuckling. 'How could you see? We were sitting round the table drinking tea.'

'Afterwards though, when you were sitting on the verandah, I saw.'

'And you said you wouldn't go to the *paneyyri*. I was very disappointed since I'd already fallen in love with your golf shirt.'

'Oh?' Christopher leaned forward eagerly. 'Which one?'

Laura leaned forward also, propping herself on one elbow. Their noses now were nearly touching and she could have kissed him. 'You fool.'

'I also have a blue one.'

'That one's not bad, I suppose, the one you're wearing.'

Christopher glanced down at himself, smoothing out creases. Suddenly Laura reached across and grabbed his hand. Without letting go, she pulled her chair around, bumping it against the table legs. Finally she managed to get close enough. She leaned her head on Christopher's well-sleeved shoulder and slowly ran her hand up his back to his neck, moving her fingers into the straight hair at the back. She felt the hair between her fingers like straw, the curved hardness of his skull pressing against her fingertips. Down on his neck again, brushing the hollow nape, her fingers like feathers, Laura was delighted when Christopher suddenly bent back his head to trap those fingers and mumbled, 'You're very nice.' He groped awkwardly with his left hand for the top of her leg. He withdrew his hand and turned, moving his right hand across her lap with more purpose. Their faces were very close, Laura holding him around the waist, the flat of her hand against his back pressing him closer, when he was on her, kissing her clumsily, opening his mouth and breathing into her most extraordinarily, with a force Laura had not anticipated.

Vasilis clunked the plates down, asking if they wanted their omelettes. 'Hurrah, the omelettes,' Laura cried. Christopher raised both arms in acknowledgement of food. '*Efharisto parapoli*.' He held out the empty beer bottle.

But Vasilis wouldn't go. 'English?' he asked Christopher, looking at Laura.

Christopher nodded, biting into the omelette.

Vasilis in his ragged black shirt stood there staring, legs apart. 'You knew her in England or out here?'

Christopher choked and again waved the empty bottle, but Vasilis wouldn't move. 'Out here,' Christopher spluttered.

Vasilis glowered at Laura. 'Is she your mistress?' Christopher shifted in his chair and reached for bread. Laura guessed the drift of the question. Christopher said, '*Ochi*.' Vasilis didn't believe him. Laura gave Vasilis a huge smile and raised the same empty bottle which faintly reflected the light at Vasilis's back. His stare was blatant as it undid her bra and unzipped her jeans. She smiled steadily with the bottle in hand, until he took it and withdrew.

'What eyebrows!' exclaimed Laura. 'Is he as Mephistophelean as he looks? Phew.'

Christopher felt feverish. He was ashamed. It was undignified, but how he wanted Laura. If he could grab her and run, perhaps . . . oh, perhaps! He swallowed the brittle omelette convulsively. When Vasilis brought salad and the wine, Christopher refilled their glasses and gulped his down, pronging a tomato to relieve the cloying oil from the omelette. He was afraid that if he looked at Laura her attractive face would freeze his courage. The fever must carry him on. Christopher ducked into another mouthful of salad.

Laura giggled. He didn't ask why.

'Are you planning to learn modern Greek?' he asked her.

'Should I?'

She was eyeing him. He took a deep breath and laid down his knife and fork. 'What should we do now?' he mumbled. 'I've finished.'

'I haven't quite.' Laura reached for the bottle and poured more into Christopher's glass. 'You're a very fast eater, or you were starving.' Laura leaned her head on his shoulder a moment, resting her hand on his leg, then withdrew her hand and cut into the rest of her omelette. His hand moved back to her thigh. She let him run his hand up the inside. He could feel the other thigh. Up, she was letting him, to the hard seam of her zip. His fingers followed the stitching down, but faltered.

He patted her knee and moved his chair away. 'Yes, you're absolutely right. His eyebrows are Mephistophelean.' Christopher pushed his chair right back, grating the chair legs against the pavement; he crossed his hands in front of him.

It was a weary effort to converse yet again with a woman. He gave Laura a faraway smile; he hoped she was wondering about his past. But she continued to eat, taking nibbly bites of omelette, her knife tapping the plate. Was she that hungry? 'He's an interesting man actually,' Christopher pressed on. 'He shows amazing courage as far as the Colonels go, quite openly speaks out against them, breaks plates, calls them evil idiots. That can be dangerous.' He talked on, willing his penis to subside and the chance of penetration yet again, oh yet again to be lost. But he would not be cheap. Sex without love was absurd.

Christopher shivered. He shook his head to unstop his left ear. With his hand pressed to it, he faced the pointlessness of the self-made businessman who pressed a girl's thighs with the length of his desire, caressed her soft breasts, licked her pink nipples . . . like a dog. Over and over again he dived into situations as bleak as in those horrible books he would read. And always he held back when he'd already gone too far, insulting the woman because he couldn't go on.

But he'd known once what it was to love and be loved back. Diana had loved him for a while. She'd made a brave man of him, until she met Saul. Short, bald Saul, with a sallow face scarred from acne, who never changed out of jeans and a grey shirt. Except when he was naked with Diana – which Diana apparently welcomed. Christopher had loved her too much to rape her. But Saul's unscrupulous New York manners ended Diana's loving him back. After all they'd done together and their promises never to part Diana had left him suddenly. The American Saul thrilled her. 'I'm sorry,' she'd scribbled on the back of his invitation to the Beagle Ball, to which he'd intended to take her in their second year; 'I'm so awfully sorry! But the truth is you're too good for me.' It was propped against the bottle of sherry on his mantel, waiting for him to get back from lunch. That adorable spiky handwriting bit into the white cardboard, 'I could never live up. Love, D.' He'd never decided what she meant. Obviously it was he and not she who couldn't live up.

He'd loved her red coat as she squeezed his arm. They met their first term at a party when he noticed the exotic hair and eyes talking to a small man in a tweed jacket. He accosted her. She laughed. Her laugh when her black hair slipped over her thin shoulder bewitched him, her deep voice promising worlds of passion. How they'd talked. They'd held hands. Everything from the first lecture in the morning they'd done together until they'd had to part at the college gates at eleven. In his rooms at New College they kneeled in front of his gas fire and recited 'East Coker' which they both knew by heart. They'd prompt each other and kiss. He needed only to touch her cheek or look into her eyes to feel complete with her. He'd thought their love was for an eternity. The sizzle of the gas fire and the hollow slamming of doors beyond where they knelt were like horns in a thick fog blowing in the invisible distance, her figure so near and clear and his. So he'd believed.

He remembered the pretty Italian shoes she used to wear, with low heels and square buckles. He'd never seen such pretty shoes again. Her ankles, her wrists, her neck were exquisite. She never met his blonde mother, who would not have liked Diana's dark looks. But he'd proposed and she'd accepted. They would have lived happily together for ever if she hadn't met Saul.

How could he ever betray such a memory? Nothing had ever measured up. He couldn't face just sex when he'd known love. He worked after that. The clock on his desk ticked in the empty room. He'd got a first.

The pain of no one left to love was still with him and made him unlovable.

Christopher's head was down. Laura had finished her omelette and drunk her wine and was ready to go. Christopher sensed her restlessness. Out of the corner of his eye he checked her face and saw her set expression, her chin up higher than was kind. Because he'd let her down. It was all so familiar.

Christopher rose to his feet. 'Well, I'd better pay, I suppose.' He found Vasilis inside arguing, those Mephistophelean eyebrows glowering at his antagonist. Vasilis slapped the table, slapping to death the other man's argument, and stood up when Christopher stepped through the door. Christopher avoided Vasilis's knowing leer, shifting his attention to the poster on the wall of a girl leaning on a rock, advertising Aroma cigarettes. He concentrated on that banality for so long that Vasilis too had to turn and look. Christopher asked him then for the bill.

CHAPTER 10

Only gradually did Jenny recover on Sunday morning from Bill's attack the night before. Slowly she could forget it, sitting in the sun with the others on the quayside at Sitia when they'd left behind the Zakro gorge and reached this pretty town. Thank goodness for the others, for Adam and Ellen and even Annabel who protected Jenny from the inevitable, miserable recriminations she and Bill would otherwise have been spitting at each other like snarling cats as they sank into hopeless gloom. The others slouched in their chairs, and the blue, red and green fishing boats bobbing in front of them calmed Jenny down. Staring idly at the decks cluttered in yellow fishing nets she reflected on her married life, and felt tension melt away as a wonderful lethargy relieved her of herself. Bill started to order lunch, distracted and anxious still, the dig on his mind. He was tortured by the responsibility. The waiter brought the wine and the pork brawn which was the speciality of Sitia. Bill let out a long sigh as he dug in with a fork.

It was strange how the others did not guess what had happened, nor have any idea how desperate she and Bill were. For so long they'd been mildly unhappy and now on the dig they hated each other because everything on the dig interfered with their married life. Even on these outings they could never go alone and abandon the others.

Why couldn't she admit that Bill's work and being out here were much more interesting than remaining all the time at Little Compton? It was her fault that they were both miserable. If she were a better person Bill would not need to attack her. She could see that clearly as she sat now with the others. And she liked to watch Bill eat. His huge frame *embraced* the pork brawn and bread and *melitzanosalata*, he was so delighted. His beautiful face came alive, his features so neat under the shock of black curls, working at them shamelessly.

She was sorry she'd made him yelp with pain. She'd hurt him last night. She did not wish to be like she was. She hated herself often and could only get over it by ridiculing Bill or anyone else who provoked this disgust with herself. It was because her mother had taught her to preserve her integrity (her mother's favourite axiom) that she couldn't tolerate other people's mistakes and flaws. To expect less of others would betray her integrity. To lose that would be the worst thing that could happen to her. So her mother always said, supporting her when anything went wrong, convincing Jenny that her perfectionism was for Jenny's own good. Her mother loved her with a mother's unselfish love which was a greater and better love than anyone else would ever have for her. When Bill lunged at her last night, Jenny was appalled – as her mother would have been. Without a 'darling' or an 'I love you' or even just a 'Jenny' he'd pulled her to him, kneaded her bottom, flattened his mouth against hers and shoved his hand up her leg to test her readiness. She wasn't a whore. Sex wasn't his right. It was due to her self-respect that she should punish him. He'd deserved to have her heel ground into his foot.

Nevertheless she wished they'd made love and not lain awake for hours seething with hate on their first weekend away after the dig had begun. Bill had pretended to fall instantly asleep, and was soon snoring. But not she. This morning she was still so peeved and tired and irritated by Bill's very existence that when he couldn't find one of his socks she told him to go barefoot, pulling the covers up over herself to blot out his stooping bulk. It was funny, looking back at it now, Bill stumbling about that airless room which reeked with the stench from the loos, groping for his socks. Such a room, however authentic, could not be idyllic. On Saturday evening when at last they reached Zakro they were thrilled by the wild beauty of the place, standing on the bluff at the sea end of the Zakro gorge, and, after a huge dinner of fish, expectant. If only when they'd shut the door against the others and were alone, Bill hadn't been so impatient!

'He should have known better,' her mother would argue. 'He wasn't thinking of *you*,' she'd insist, returning Jenny to her unbreached integrity. Jenny was trapped in her mother's support. But she should be brave and risk forfeiting it. It was ridiculous that at her age she should still be so influenced by a parent. She knew her mother was a snob and prejudiced in all sorts of ways. Her mother and Bill would have fierce arguments which Bill seemed actually to enjoy. Jenny's father would withdraw behind the newspaper. Sentimentality was one of her mother's favourite faults. And hypocrisy. So deeply had these sins sunk into Jenny's mind that when she and Bill were first married she accused Bill of being both a hypocrite and sentimental when he called her 'darling'. Soon he gave it up and now she wished he hadn't. But she'd been so fierce about it at the time that he'd naturally decided he was displeasing her when he showed affection.

Damn her mother. Her mother was wrecking her life. And how she flirted with Bill! She'd swing her leg to show off her thin ankles when Bill was in the room. She'd egg Bill on to make some pronouncement that she could hurl back at him, and accuse him of 'dishonesty' or 'arrogance' or 'cruelty'. Jenny wished her mother would let her hair be naturally grey, but her mother, pleased with herself as always, said the blue rinse showed the world she knew her hair was grey. It was honest of her!

'I was just thinking about my mother,' Jenny murmured. 'I can't think how my father puts up with her.' The pork brawn raised them from their private reveries as they leaned across the table to prong the jellied concoction, and washed it down with the rosé wine which was better than what they drank every day in the village. Adam looked up, surprised.

'Do you find your mother difficult?' he asked.

'I suppose I do.' The waiter set plates of red mullet in front of each of them.

'Jenny's mother's quite a tartar,' Bill agreed.

'But you like her?' Jenny asked.

Bill's eyes jerked from his fish to the Greek family at the next table who had already been served salad although they'd come later. He looked at his watch. Jenny's question didn't sink in.

'My mother's impossible,' Adam started. He kept his eyes down to avoid anyone's look, especially Jenny's since his mother was her friend. Slowly he began to describe his mother's rules about dishcloths and buckets and dustpans and hangers, and the hysteria and headaches that followed if her rules were broken. 'She was pretty twenty years ago before buckets mattered,' he explained with such dry tolerance that the rest of them giggled. And he went on and on until they were all laughing so much they couldn't eat. Even Bill forgot the missing salad.

'Sometimes a guest forgets that the pink bucket is not for organic but *in*-organic rubbish, which is terrible when my mother finds potato peel in it. It's too much for her, in fact. She won't speak to the guest who's done it. She goes to her room with a headache until my father has cleaned out the bucket. It's difficult for my father and me to explain to the guests why my mother has gone to lie down because they probably wouldn't understand and might think that my mother is slightly mad, which of course she is, but it would be disloyal to admit it and my father is very loyal, so nothing is ever said. I find newspapers for the guests to read until my dear mother has recovered. When my father and I hear the bedroom door open I quickly bring up a fresh topic of conversation like modern architecture so that the guests won't hear the thud of the buckets and the give-away commotion of their wire handles coming from the kitchen as Mum spends her little time with them before she can join the rest of us in the sitting room. She must be sure, you see, that all is normal again. I think she fears that her life will break up and lose its meaning if things aren't normal and the buckets are left like some evil thing to gnaw away at her sanity. My poor mother. I try to remember that she's behaving in a quite human way.'

Jenny exploded with laughter, shocking the others. But it was her friend Harriet that Adam was describing, her practical, forthright friend whose practicality and forthrightness made Jenny feel so inadequate and grateful. Harriet always knew what to do. She would decide for Jenny and act, like the other day when she took Jenny to the ironmongers in Chipping Campden to buy a greenhouse heater because Jenny could no longer afford to send sheets to the laundry. Dear Adam. How funny he was being! It was a relief to laugh. Jenny saw how Bill's worried forehead was quite smoothed out. He was laughing almost as much as she. She had no idea Adam could be so funny. She wanted to hug the dear boy who looked meek and mischievous leaning back in his chair, his narrow head beautiful. She had never seen before how the thin mouth and short nose and brown eyes were all of a piece.

On the way back Jenny made one resolution after another for the following week. She would not mind about the shopping, not complain that people were using too much loo paper, not nag about the washing-up. If Margaret's food was not appreciated it didn't matter. Adam had inspired her to try and be a better person. For the first time in weeks she wanted Bill to make love to her. Both Bill and Laura deserved her apology. She was about to do something her mother would not approve of. Jenny's eyes filled with tears as she thought how Bill needed to sleep with other women because he was married to someone like herself.

CHAPTER II

As the morning wore on Laura longed for her nap. She'd been put to sweeping Susan's *dromos* which was now clear right up to the blocking stone of a tomb chamber, and Christopher wanted to take photographs before the boulder was pulled away. Susan was excited. She showed Laura what to do impatiently, anxious to finish drawing her final section of the *dromos* with the blocking stone in place. Laura was to sweep the boulder, the earth round it and the floor of the long *dromos* until she reached the baulk. She was to make sure the bristle marks of the broom went always in the same direction. As she stood at the level of the entrance to the tomb chamber, the blocking stone was as tall as Laura. Soon her face was white with the dust she swept out of the jagged crannies of the rock.

Susan shouted to Laura to move a minute. She tossed her a tape to hold up against the stone and read out where it crossed the other tape which had been stretched across the *dromos*. 'He'll be over in a minute!' Susan chivvied her. 'He can fuck off!' Laura whispered. She was so mortified by what hadn't happened on Saturday night that her day on the beach on Sunday had been ruined. Why did Christopher talk about Vasilis and the Greek Colonels a second after his hand had been groping her leg? He was wrong if he thought she was an easy lay. She had high standards actually. Truth and beauty, hope and charity meant everything. If she hadn't found his bedroom littered with filthy novels she wouldn't have thought for a minute that sex ever crossed his mind.

She'd tried in the game of charades on Sunday evening to make a joke of their Saturday night debacle. Christopher laughed when she collapsed into his lap. But at breakfast Christopher didn't even look up. She was back to being a nobody. When they reached the cemetery hill he had set her to sieving all the loose earth from his robbed tomb, afraid of missing something. And now Susan made her sweep.

Susan finished her measuring. Laura carried on. She moved slowly away from the blocking stone, up the *dromos*, the dust swirling about her in the narrow ditch. They were a heartless lot, these archaeologists. The air was heavy; leaden clouds dulled even the peaks of Dikte, which looked momentously close. The landscape was as changed as Christopher, clumps of green thistle and sticky cistus and scrub pine brown and soft, the valley sullen and bog-like in a strange Irish light. Her walk up the valley in the dawn cool had restored her a little after no sleep and dispelled some of her irritation with Annabel for snoring all night. But Christopher's silence hurt. He made her feel at fault.

Christopher shouted for Andonis. Laura heard a clatter and took off her scarf and dusted herself down. Christopher and Andonis were trying to make a stepladder straddle the *dromos* so that Christopher would get the photo of the blocking stone straight on. It wouldn't. Adam came over to help. They shoved a board across to support the legs of the ladder. Andonis was shouting, the excitement mounting. '*Chryso*' was the word Laura heard repeated by all of them while Christopher adjusted his camera. The boy Manolis raised his arms to dance, after kicking pebbles away to clear the ground. Andonis admonished him, and Manolis walked off in a sulk.

An ungainly figure in khaki trousers and blue denim shirt as he swayed at the top of the rickety ladder, Christopher took shot after shot, the camera itself held with steady ease as he checked every time on the light and the angle. Laura squinted up at him, wondering what he'd do if he fell off.

When he finally finished, the ladder was collapsed and Andonis clattered back with it to its place under the olive tree. The point of Laura's sweeping was over since everyone now jumped down into the *dromos* and messed up her brush strokes. Christopher asked Susan over the heads of Adam and the workmen, 'You think the *dromos* burial was later?'

'L.M. IIIC.'

Christopher looked sceptical. 'We found nothing that late last year.' He shoved his hands into his pockets.

Andonis, Manolis and Adam pulled unsuccessfully at the stone. Andonis fetched his shovel and tackled the stone from above. Then he ran for his pick. Laura wished she could understand these L.M.s and M.M.s and IIIC.s which the archaeologists bandied about. It was like her talking about white sauces and blanching. But she wasn't in a kitchen up here on this hill; she was standing *perhaps* over dead people's treasures and bones. Susan obviously hoped to find gold. Christopher appeared the least excited.

With the pick Andonis pulled like a charioteer. His thick arms covered in grey hair looked improbable.

'So you think you might find several periods of use in the chamber?' Christopher went on to Susan.

'*Oopah*,' shouted Manolis. The stone was moving. Andonis strained at the pick.

Now Christopher was down in there with Adam and Manolis. Suddenly with a dull 'wumph' the stone fell onto its back.

All six crouched up to the hole and peered in, hushed by the fact that it was an unrobbed tomb indeed. Only two cups had spilled out. Something round lay in the dirt further back. The earth inside was as it had been left three thousand years ago when the stone was pushed against the entrance for the last time.

Christopher wanted to take more photos; he shooed everyone out of the *dromos*. The ladder came clattering back. He also took close-ups. Nothing could be clearer or more certain, he announced, than that this was an unrobbed, unexcavated tomb. He sounded surprised. When he finished photographing it was time for the mid-morning break.

It was Jenny's 'Who's washing up?' that broke the silence. Most had finished their lunch. Jack pushed back his chair and started to collect plates. The clatter as he stacked made a dismal end to the meal. Laura laid her hand on Margaret's shoulder as she left and thanked her for the delicious meat loaf. It had looked like an exotic fish, thin gills of red peppers decorating its great length. Margaret had also produced dishes of leeks in a cheese sauce and boiled potatoes, arranging them down the silent table.

Of course Laura had eaten too much and once outside she walked at a run, her hand pressed to her straining waistline. She was eager to tuck herself up on her camp bed before Annabel arrived. She dreaded the shuffle of Annabel's canvas shoes moving purposefully about. But Laura woke up several hours later to an empty room. Annabel's sheet and blanket were still in a crumpled mess on her bed, her orange nightie on the dusty floor.

Laura pushed her hands under her pillow and stared languidly at the stains on the ceiling as she considered Susan's lust for gold. Susan thought there would be gold deeper down with an earlier burial. She had insisted on this when Christopher discussed what they would be finding. Before they lifted anything out, it had had to be drawn. With amazing speed Susan drew the three jugs and the stone bowl, her hand moving deftly from point to point on the graph paper as she and Christopher argued. By then they'd already pulled out many cups with smooth flat handles which surprised Laura by their lightness when she picked one up.

But did Christopher really not care if they found gold? Could he just be interested in the architecture of the doorway into the tomb chamber, and its alignment with the *dromos* (that damn ditch she'd had to sweep)? His professionalism couldn't be that lacklustre! People rushed across the whole continent of North America to find gold. Could Christopher be that inhuman? Oh, really, Laura chuckled to herself, stretching her legs right out until she could curl her toes over the edge of the stiff canvas. That felt better. Her feet ached.

Laura's thinking veered back to her absent roommate and she wondered if Annabel was really interested in archaeology like Christopher and Bill and most of the others. She turned on to her side and shut her eyes, imagining now how it would be if John lay beside her. Nice. She pulled up her knees and clenched her teeth until the longing passed. Jerking herself back into a thinking position, she glared at the same green stain creeping down the wall over her head and admitted to herself, pushing the hair back from her face, that she could only lose herself in the arms of a man. Was that an awful thing to have to admit? At least she could face up to the truth. No Annabel would admit as much in a lifetime, laced in lies and deceptions and brash pretensions, like that silly orange nightgown in a heap on the floor. Who did she think was going to see her in that?

CHAPTER 12

Since Christopher was so keen to have his way with the tombs, Bill decided to let him. If there was trouble he could cope. This suited Bill. Christopher's cemetery hill might be rich in treasure but Bill's tower, if that's what it turned out to be, at the western edge of the Kallithea site would be an amazing find, quite in another class, proving wrong the assumption that the Minoans had no need to defend themselves. Bill had never gone along with such a naive idea. How could there have been a Pax Minoica all those thousands of years ago?

Four beautiful courses of large, roughly squared stones were already uncovered and there was no telling how far down it would go. At the southeast corner of the trench the wall continued east at a ninety-degree angle, a good two feet thick. When they reached the levelling course they would lay out another trench to the south to see if the other corner of a tower existed. The use of this strong wall was long-lived; already they'd found three Late Minoan levels and it seemed from the shiny black pottery turning up now that they'd reached Middle Minoan. How exciting if they were finding the abutment of a Middle Minoan fortified village, say, or a castle. How furious Lambson in London would be if he were to find out that the Minoans had not been peace-loving flower people. The greatest art, Bill felt anyway, was a poignant expression of longing in brutal, insecure circumstances. The Minoans painted monkeys and birds on their walls, and made pottery as thin as glass and as elaborately decorated as a tapestry. Such aesthetic achievements were out of the question in a wet blanket of a world, under a Swedish socialist government. Lambson was a fool and Bill was going to prove it. The Minoans were as warlike as the Mycenaeans and the Germans and the Romans and the British and any other people who tried to survive on this earth. Thousands of years ago the bare hills of Crete might have been wooded and more lush, but the Minoans must have worked even harder than the Cretans worked now to survive. Pirates and robber gangs and mainlands with expansionist policies must have been a constant threat.

The swim had cleared Bill's head. The disagreement with Christopher the night before no longer niggled him. He pulled his towel tighter and buttoned up his shirt, slopping along in his unlaced tennis shoes past the cafés. Ahead, sitting alone at one of the tables, he saw Mary Elizabeth sipping a cup of coffee with her notebook propped open on her lap. He suggested he join her in a minute for a drink. She had changed into another pale, ironed blouse and a pretty cotton skirt. All morning up on the site she'd worn a sleeveless T-shirt, her neat breasts bra-less under the white cotton as she reached the fourth tread of her gypsum staircase. He'd asked her right at the start if she weren't cold, noticing goose pimples peppering her bare arms, but she'd squinted her eyes in a most lovely, girlish way as she looked up into his face, too excited about the fig she'd just found on the third step to feel the dawn chill.

The fig was unimportant. Badger, who was her pickman, made a great joke of it with Bill, which Mary Elizabeth was too taken up with ancient subsistence problems to appreciate. Bill had shouted down to his old teacher, 'The Minoans ate figs!' and his friend shouted back, 'The Minoans were human beings!' Badger went on to list the other things the Minoans might have eaten if they were human beings: 'Spinach pie, cheese pie, apple pie, nuts, raki. My wife is a Minoan!' he'd chanted, wagging his head. Mary Elizabeth's silky hair slipped across her cheek.

Bill ordered her a beer and himself a carafe of raki after he'd pulled on his flannel trousers and sandals and shaved quickly at the sink in the kitchen. A wind swirled twigs and bits of paper past their legs. It unsettled the street, but Bill was enjoying himself too much to mind.

'How do you explain their mint condition?' he asked his beautiful companion, with eyes fixed on her young neck. 'Any idea?'

'L.M. I.B.'

'So you think the staircase was never used. How deep was your destruction level?'

'Two metres.'

How Bill wished it were Mary Elizabeth and not Ellen who was supervising the tower. It was the tower which preoccupied him but to discuss it with Ellen would not be half as much fun as this. With Ellen he had to suffer a flood of interpretation before he could begin any kind of discussion: 'You can see that this wall goes with this floor and over here you see where something heavy fell and was removed much later – the earth is packed down and darker. The imprint of something large and heavy is clear,' she'd fling at him before he'd had time to look. Who wouldn't prefer Mary Elizabeth's short answers? He could then give his view and work out his own ideas aloud. 'Over here I've noted a reddish colour and it crumbles easily so there must have been a small fire. It's confined to this one spot, so the fire couldn't have been a destruction. Nevertheless it's certain there was a fire, we've reached a floor and the pottery is local but clearly M.M. II.' Bill sighed and gulped down the raki, which warmed him. Perhaps if Ellen were prettier, she wouldn't be so aggressive. She was the sort of person who couldn't learn that certainty in archaeology was absurd.

'What do you make of our tower?' Bill asked Mary Elizabeth, touching her arm.

'Darling John,

'I suppose those dahlias you sent me are properly wilted by now. I forgot to throw them away. Never mind. Here is a long way from Fulham and when I get home I'll open the windows and let that "sweet" London air dispel the sour odour of my sour past. And start again. Isn't that right, John? Have I learned my lesson?

'You wouldn't believe my bedroom here which I share with a gaudy, middle-aged Pimlico antique dealer. She thinks she's God's gift to archaeology, and gives the second-in-command called Christopher Bendick her full attention. She wears a frilly orange nightie. No Christopher as yet, though, has stormed our citadel. News of the nightie hasn't reached him! Through the doorway we have a view of the flat roofs of other native Cretan homes where bent old women walk about, silhouetted against a pink sky. Today there's quite a wind which rings in the rusty cables. Ancient chickens cluck for no apparent reason since their roost is well off the ground. They are a sad lot, one has a bald neck, another keels over with a broken leg, another is blind. Someone looks after them because the broken-leg one is always being returned to the pile of sticks when I'm not looking.

'Are you asking, am I happy? It's difficult to say. I do what I'm told. Jenny has taken to swatting flies. Maybe it's the policeman on her mind who's been frightening her in town when he won't let her park anywhere and wears a gun to prove it. Lunch was bad with her wopping them, wham, just as you were about to enjoy another mouthful of chicken curry. Bill and Christopher had an argument at dinner last night. You've never met Bill, have you? He's a big man, very tall with a huge head. Only dabs of grey hair, lucky man. He smokes and when he talks, if he's laying down the law (which he often is), he wags his head in that false, upper-class English way. Only false because the wagging suggests such and such *must* be said because everyone *agrees* it's true. You don't do it, nor does Christopher Bendick actually.

'But the Argument of last night. It was Achilles and Agamemnon at it. Bill said Christopher was a fool not to have put Andonis or Manolis to guard Susan's tomb through the night (I'll tell you about Susan's tomb in a minute). Christopher stared at his bowl of soup in silence, which enraged Bill. "Well, aren't you?" Bill shouted, and slapped the table. "We're in a den of thieves here and you leave this tomb unguarded for anyone to go and take whatever they like!" Softly Christopher reminded Bill that it was nearly cleared and that if a thief did have a go, he wouldn't find much. Susan piped up then, unwisely, I think, with, "We could still find gold. You never know." Again Bill slapped the table. "It's utterly irresponsible of you," he pronounced, wagging his head. "I don't think it is," Christopher retorted, his voice also rising. "And how can you possibly say that?" Bill asked. All of us had stopped eating, only Margaret the cook fretting about the soup getting cold. Christopher and Bill sat across from one another at the street-door end of the table. Christopher took a mouthful of soup, it was brilliant. The way he frowned then, he might have been deliberating over whether the soup tasted more of leek or carrot. I was *very* impressed. Almost languidly he then answered, "Because I did the responsible thing in considering the health and welfare of my workmen. I was determined that Andonis and Manolis should have a good night's sleep. Andonis is quite old and Manolis is very young. And they have both been working hard. There is no gold at the bottom of that tomb." Bill leaned against the table, quite beside himself. "How do you know?" Inevitably the unwise Susan pipes up again with "He doesn't." And that started everyone laughing. Bill hung over the table like a fighting politician. Adam Loveday (the architect who's a neighbour of Bill's and Jenny's in Little Compton, a disconnected youth but all right *au fond*) *yelled* down the table at Susan, "Your stars can tell you if there's gold or not. Ask them!" and Susan's retort was, "Stars don't predict about *things*." Funny how that deflated Adam, woof, just like that. Hunched over his soup bowl, he mumbled something like, "Ask them then if *you're* going to find gold." Susan has some hold on the boy. He left soon afterwards, banging the door behind him (he's a great banger of doors), more or less in Christopher's ear since he was sitting right there. Christopher flinched and Annabel next to him complained about Adam's manners.

'Bill still hung over the table. "I am responsible to the Greeks that this excavation is run properly and that every precaution is taken that's needed. Are you listening to me, Christopher? Listen to me, will you?" he shouted, poor man.

'So you see. Life's not dull. Today I wrote out labels for all the cups and jugs they pulled out of Susan's tomb. That was giving me *some* responsibility since I could have written the wrong thing on the label. They found more and more today, including a skull with a stunning set of teeth and a bronze dagger, but the going is incredibly slow; the burial chamber has become so tiny that only the boy Manolis can fit in there with his pick and of course they're always stopping him. That poor Manolis has to keep crawling out so that Susan and Christopher can measure and draw and photograph while the rest of us stand about. I'm not imagining myself either in a few years as a skull with super teeth while I wait around; my teeth aren't that good. And the Minoans were just *so* many thousands of years ago, my imagination can't stretch that far. What you, John, might be doing at that very moment is more my line of thinking on that windy hill with the mountains behind and the sea in front and the lush strip of orange groves down below! They'd thought, you see, when they started work this morning that they'd about reached the bottom of the tomb. And they hadn't. They've found another, earlier burial with a whole extra set of chattels. Were you just then biting into a steak *au poivre* with your mates?

'The weather has changed. Suddenly you see the gold top of a hill when a frail, pre-Raphaelite ray of sun pierces the grey clouds. The wind's got up and it's much cooler. Now I'm sitting on my camp bed "having a nap". Christopher over our second breakfast under the olive tree this morning invited Andonis and Manolis to have supper with us this evening. I'll not understand much but I like looking at the faces of those two, Andonis's with the deep creases for crow's feet which express an ancient kindness despite his sceptical remarks, and Manolis with his huge brown eyes and head like Alexander the Great's. The boy is very clever, Christopher thinks – on the way to becoming "a good archaeologist". Great praise, that, from a Christopher Bendick type, who is the aloof sort of Englishman who shrouds his thoughts in handsome mystery and is punctilious about paying the bill. Bill and Christopher make an odd pair to be running this whole affair, they're so *equal* in authority and reputation apparently. Whatever do the Greeks make of it, I wonder, because though Bill is director, Christopher does a lot of the directing.

'God, why aren't you here! You remember? We'd seen *Who's Afraid of Virginia Woolf*. In your most courtly fashion you'd written me a note asking me out to the theatre and dinner. I sent you a telegram back, to your office of course. And what a thing to have gone to, our first evening out! No fodder for sex, no preamble to romance. In the intervals we spat barbed witticisms at each other, infected by that bitch and her sod of a husband. In the third interval we walked out into the street to cool off. It was still light, people were milling about in the summery evening. There was the smell of tar and the hiss of the machine where they were repairing the street across from us. You told me I had good legs. I cuffed you on the shin and told you to watch your step. And the fourth act was not so bad. Afterwards you took my arm and guided me out of the theatre. The hiss of the tar machine and the low fire under it burning into the street attracted us – we crossed over. Your arm slipped round my waist and then we were wildly kissing each other, clinging to one another at the dark edge of that hissing fire. Was it the play? We'd emerged from a spectacle of hell to the sweet smell of tar. Was it the tar?

You hailed a taxi and I gave my address and we began to make meltingly ecstatic love as the taxi idled at lights and accelerated, knocking its comfortable clunk clunk clunk under us, driving us to Fulham. Do you remember how you tore off my clothes? Although it was the first time you'd been to my house, you had your hand between my legs as if you owned me, propelling me to the sofa with demoniacal energy. And we laughed then, when you had me on my back and you were inside me and we were at it like two old pros. Dinner was much, much later at my kitchen table. You never cancelled the restaurant. It was too late by then, and all I had in the refrigerator were some eggs and a bottle of Muscadet.

'Come out. Let's not say it's over. Not yet. Another year at least. We've never been out of England together, we've never risked so much before, I suppose. But couldn't we now?'

Laura scratched a large illegible 'Laura' that trailed off the edge of the paper into her lap. Her face was wet with tears; she sniffed and had to feel under her pillow for a Kleenex. But sobs welled up. She clutched her face. When she looked up to wipe tears away, she wailed into the empty room. This was so silly, she told herself. But she couldn't stop.

Something heavy rolled overhead, a little bit and then more and then a long loud roll back the whole length of the roof. The wind blew sticks against the door. The chickens clucked in ruffled bursts outside.

Laura calmed down. Bleary-eyed, she scanned the sheets of scrawl scattered on the bed. She couldn't send any of it. She'd said goodbye to John at the door. She'd waved the car away, that old grey BMW littered in the back with ugly, broken-off bits of toys and an odd Wellington boot. With a big smile on her face she kicked a foot into the air as he pulled away from the curb. She meant to look jubilant in his rear-view mirror. No more goings-over and 'I can'ts' with a long, after-dinner whisky in hand. Laura blew her nose. Thank goodness Annabel hadn't come in and found her weeping. It might have surprised her to find the beautiful, confident and witty Laura, Jenny's best friend (and Bill's first mistress), in a wet huddle on the bed.

Slowly Laura put the pages of the letter together; some bits she glanced at, pleased by her 'frail, pre-Raphaelite ray of sun'. John would know exactly what she meant, they'd so often looked at pictures together. A mosey down Cork Street had been a favourite pastime after lunch, when Fulham was too far to go before John had to be back at the office. Once he'd spotted his wife stepping into Waddington's, a tall, thin dark woman like herself. He'd pushed her back into the Redfern and crossed the road to spy on his wife. Laura had hated John's giggle when he rejoined her. The correct thing would have been to apologise, but he didn't. He liked the Boy Scout stalking part of the whole affair which staunched shame in him. Sitting now in this horrible room in the semi-dark with the wind tearing everything apart outside, Laura felt used. She had good reason to ask if John *felt* anything about anyone. The use people were to John was all that mattered, and her use was excitement. She turned him on. It was her legs he'd first noticed when they were introduced by June Morrell at the Morrells' Sunday brunch. They'd discussed orange marmalade because he'd just made some which hadn't turned out well, and she'd made him laugh with her wit about the bitter fruits of his labours. But his mind was on her legs. He approved of her looks which were very like his wife's. She'd argued that grapefruit marmalade was better anyway. She was recovering from Francis's alcoholic breakdown and didn't care much what she talked about. She was glad to be in a room full of people, away from her lonely little house in Fulham which was her settlement in the divorce. But Fulham was not where she wanted, at forty, to end her days. Fulham was a dead weight of loose roof tiles and weak pipes. She had to accept it because she'd have been a fool not to, but she hated living there alone. An image of that unemptied coffeepot and her dirty mug on the white enamel table got her off the bed. She made a wad of the letter and stuffed it under her pile of T-shirts in the suitcase.

Dark, fleeting clouds battered the moon, dishevelling its fullness in the wind. The clatter of tin cans blown down the street made Laura hurry to get inside. She had put on her party dress which was bright red and plunging, and tight. She had put on earrings, large, flimsy circles of gold which she clapped her hands to in the wind. She'd jammed pins in her hair to keep it piled up on top. Her lipstick she'd retrieved from the bottom of her handbag. This was the first time she'd worn it since she arrived in Crete. It was her weapon against the wind and all the others inside sitting around the table waiting for supper.

Only Christopher looked up when Laura stumbled through the door. His features froze, or seemed to in the fitful light; perhaps it was only that he didn't move, the full expanse of his naked face exposed to her lipstick and bare neck. He'd never looked more serious! Laura made sure the door was shut to keep out dust and squeezed past. Then on an impulse, the serious face too inviting to begrudge, she turned back, laid a hand on his shoulder and whispered, 'I couldn't let the men down, darling,' chuckling lewdly into his ear.

Annabel at Christopher's other ear heard 'darling'. She took up her knife and tapped it on the table.

Jenny berated Laura for risking her best dress in such a wind. Laura sank down into the empty chair beside her. 'Oh well, I had to,' she murmured and Jenny, in the same drab waistless dress she'd worn all day, with a cardigan draped over her shoulders, regretted that she herself hadn't put something else on. 'Christopher didn't tell me he'd invited Andonis and that young one to supper, or I would have dressed up too,' she complained. Laura laid a hand on Jenny's. Jenny's eyes filled with tears as she wished she weren't herself.

Something clattered and banged in the kitchen. Margaret said it was a loose shutter when she brought in the first plate of food. She'd made a Greek *pastitsio*, two huge dishes of it, the tops decorated in an un-Greek way with a pretty scattering of finely chopped parsley. One was put in front of Andonis and Manolis by Mary Elizabeth, who was helping Margaret bring things in. More and more food covered the table: spinach pie, bowls of taramasalata, garlic bread arranged on several plates, salad, cheese and olives. Adam groaned, Susan laughed. Usually supper was a light meal, soup and cheese and wine, not this procession of different dishes. While the shutter banged and a tin can scraped and clattered outside, they tucked in. Andonis wished everyone health, Manolis did the same, Bill responded in his loud Greek, and Christopher thumped the table with his glass in Cretan style before quaffing the wine.

Adam had been drinking for several hours in a café. Now suddenly he stood up and started a speech to welcome Andonis and Manolis to 'headquarters'. He held out his glass and squinted down at the oil lamp with the damaged expression of someone who's just been chopping onions. Forgetting why he was on his feet, he fell back into his chair and stared at his food. Andonis, a few places away, called out to him that he was a 'good lad' and reached across to clink Adam's glass.

Margaret had made far too much food; everyone was too tired to eat and more than half of the *pastitsio* was left. She walked back into the kitchen with the dishes, staring down sadly at the undisturbed surfaces of chopped parsley. Jenny called after her that it would do fine for lunch the next day, but Margaret had already planned what they'd eat the next day. In the afternoon when she'd finished with the *pastitsio* and taken it to the village oven, she'd straightaway moved on to the fettuccine, doggedly rolling it out on the wobbly trestle table, stringing up a line across the dark kitchen. It was too bad about the *pastitsio* – she threw it away into the plastic bucket and washed the bowls before anyone could see.

Adam was supposed to wash up. Thinking she'd better remind him, Margaret noticed when she tapped him on the shoulder that he'd not touched his food. The undisturbed heap of spinach pie and *pastitsio* was an insult. She jiggled the back of his chair and told him to eat *something* –at which point he squinted up at her lank hair which hung over him and lunged back in his chair, splattering her feet and the floor with vomit. Margaret cried out. Laura and Annabel bumped into one another in their rush for the bucket under the kitchen sink.

It looked as if Margaret too would be sick, the way she gripped the chair. Adam laid his head down on the table. Jenny pressed the kitchen roll on him and told him to sip a glass of water.

Annabel ordered Adam out of her way. By now she was down on her haunches in her khaki riding breeches asserting her sanity and competence. Edward volunteered to help Adam to bed. Out in the gale Adam leaned on Edward's shoulder, the door left open long enough for two empty cigarette packets and a fluffed up, live chicken to blow in. Andonis, who had raised his glass to Adam as he left, grabbed the chicken by a leg and tossed it back out into the night, shutting the door.

Laura poured out more wine and suggested a game of backgammon. She raised the jug in the air and challenged Andonis. Often she'd heard Susan and Christopher up on the cemetery hill brag how the clumsy shoveller was a demon at *tavli*.

Andonis shrugged and drank down his wine. He ordered Manolis to go and fetch a board from the café. How was he to view such behaviour? Was it an insult? Was it wrong? The red dress was all right. She had a good bosom and good legs. The legs he knew well from up on the hill when sometimes she'd worn a skirt. She was not a good woman but she was a foreigner and had spirit. She had pleased many men, a fact which both amused and embarrassed Andonis: thinking of her as a foreigner he was amused; thinking of her as a woman he was embarrassed. At this moment he was mostly embarrassed, since it was a shame to play backgammon with a woman.

But the woman had spirit and he liked spirit. Andonis grabbed the jug from Laura and refilled his glass.

Annabel said she'd empty the bucket outside, and gripped the handle as she marched down the room. Conscious of her consideration of everyone else as she pulled the door behind her, she was glad to be out in the wild night, disgusted by Laura's behaviour and by all the others who tolerated such showing off. Jack giggled and in a loud whisper to Susan said he hoped Annabel didn't get it in her face.

With her feet cleaned of Adam's vomit, Margaret went about the kitchen doing what there was still to do before she could go to bed. She hoped someone would come soon to help. Laura's silly behaviour didn't impress her very much one way or the other since Laura looked old and desperate in that dress.

'*Deka!*' shouted Andonis. He handed the dice to Laura who hadn't seen what he'd thrown. 'Throw!' he ordered her. 'Hurry!' She threw in her English, underhand way. The dice rolled slowly across the board. Andonis snatched them up and said it was eight. He pitched the dice. Laura clapped her hand on them before he could snatch them back. 'Seven,' she called. She held up seven fingers. Andonis shot his counters forward. He waved her to hurry up. Laura counted out his move, the dice safe in her hand. Andonis turned to Christopher to complain, but Christopher was watching Laura as she leaned over the board. She frowned slightly, concentrating with rigid calm on the game. She thrilled Christopher. He warned Andonis to watch out, Laura was a clever woman.

Laura grew faster, but she paused whenever it was her turn to throw. She moved her counters quickly. When she handed the dice back to Andonis, she would provoke him with her smile. He threw, snatched, moved and thumped the dice down for her to take her turn. Jack and

Susan left. Jenny watched for a time, then left. Bill's head was tilted back with his mouth open, his torso bent backwards over a café chair; he looked like *rigor mortis* after the agony with his hands nicely folded in his lap. Ellen and Margaret averted their faces from such undignified unconsciousness and hurried out into the night. But just after they'd gone there was a knock, loud and meant, not something blown against the door by the wind. Christopher got up to see.

Vasilis brought with him a short thickset young man in a white shirt with dishevelled orange hair. The wind had worked up to a dull roar, the door every time harder to push shut. Even Vasilis looked blown, his dark mat of hair twisted although his black T-shirt was as tight as ever. He shouted to Bill that he had brought him his cousin. Bill woke up with a snort.

Andonis shifted in his chair, ashamed to be found playing with Laura. 'That wasn't a four, it was a five,' Laura insisted. She jabbed her finger at the board until Andonis moved back his counter.

Vasilis led his cousin down to where Bill now leaned on the table rubbing his eyes. He pushed the cousin between them. 'He wants work. He works like the demon. Give him work. He's a good chap.' Bill stood up and pushed his hands deep into his pockets.

'I won!' shouted Laura from the other end of the room. She threw up her arms. Vasilis's cousin looked round. His eyes took in the red lips and the red bosom lifting in triumph. He moved down to this amazing woman. In English he asked Laura if she was a 'champion'.

Laura gave the orange-headed stranger a rapturous smile. 'He tried to cheat and I caught him out!' But Andonis splayed out his fingers to dismiss Laura's triumph as a freak. The redhead laughed. Pulling up another chair he challenged Andonis to a game.

A restless aftermath beset Laura. She looked for the wine jug. When she found it upside down by the sink she refilled it from the demijohn on the floor and set clean glasses on the table. But Bill refused, as did Christopher; only the redhead and Vasilis would have some. She filled their glasses and forced a full one on Andonis who swilled it down, thrusting the empty glass back at her without a word. Christopher and Bill were both discussing with Vasilis whether to use his cousin. Eventually Christopher admitted he needed someone to take off the surface level of a new trench. Vasilis slapped his cousin on the back and announced that all was arranged.

Laura was in no mood to leave. The thought of Annabel already snoring in the other bed spurred her to rejoin the game. She watched the hairy wrist of the redhead throw dice and push counters. The point of the game faded. The ratatatat and the movements of hands in the wavering shadows dulled her restlessness. She leaned her head on one hand and drifted into a wide-awake stupor. A hollow thumping started overhead.

Christopher touched Laura on the shoulder. 'We should all go to bed.' Laura gave him a lazy nod.

CHAPTER 13

Christopher stood with his trousers undone in the dark room, baffled by himself. Laura had been silly. When she arrived dressed like a harlot and whispered that she 'couldn't let the men down, darling', he was shocked. That much cleavage at a simple supper for two of the workmen? Outlandish, too, were the flimsy dangles. It was embarrassing.

So embarrassing that he had wanted to whisk her away before she did anything else wrong and rescue her from the others' contempt.

She spurned his efforts. She showed no shame. Instead she forced Andonis to play backgammon. Because she was English and *o Klystopha's* colleague, Andonis couldn't refuse, although for a Cretan to play with a woman was difficult. She was tactless and ignorant. And he had pretended to approve. He even urged her on. When he warned Andonis that she was a clever woman, he betrayed his old friend for her sake.

At least she should have shown tact enough not to win. She won! And he had laughed. That was the worst thing he did all evening. He prayed Andonis understood why he'd laughed. It wasn't at Andonis's humiliation. God forbid. He'd laughed only because it was so ridiculous that Laura in that silly dress won against someone they all knew was a champion.

Loose tiles clattered overhead. The pine tree whirred outside his window. Was Laura still at headquarters draped over a chair? Had she stayed on? He wished he could forget that ridiculous red dress and those earrings which made her beauty so pathetic. There was something wrong with Laura. Why last Saturday did he grapple with her like a teenager, and under Vasilis's nose run his hand up her thigh? What had made him forget himself? She'd tried to seduce him. She was obsessed with sex. Which might explain this evening. On Saturday he'd backed off. She wouldn't try him again. Perhaps she'd divined that he was still a virgin.

Who was she after? Bill? Christopher dropped his trousers. He felt happier than a moment ago. The thought of Laura and Bill rolling about in bed amused him, Bill peering over Laura's bare shoulder to look at his watch. Laura would have trouble holding him down for long.

She didn't even know the man's name. Yes she did. Michaelis. That was it. Michaelis.

She yawned, chuckling to herself, stopping abruptly to see if Annabel was in the other bed still. She wasn't. Laura stretched her legs as far as they would go. She stretched her arms up over her head. She yawned with abandon now, to the very limit. Dropping her arms limply either side of her, she laughed and confessed that the red dress was a bit the worse for wear. But why not? Why ever not?

The heavy thing rolled again overhead. A short roll, then stopped, more rolling and on, then the other way a long continuous, headlong grinding roll. Laura braced herself for the thump when it toppled off the edge. But it didn't; miraculously it stopped and rolled a little way back again. How very odd. The wind was still blowing, clammy weather as it had been yesterday, she expected, low milky clouds shrouding everything in a Day of Judgement oppressiveness.

Safe and sound in this dark room, Laura supposed all the same that she'd better get up.

Suddenly the 'better' struck her. A work day. What time was it? She threw back the covers and was up, groping under the chair for her alarm clock. Stabs of fright killed her calm. Bill would be furious. And what would Christopher say? Would he guess what had happened? Eight o'clock. There was just enough light to see by. Until she had put on her T-shirt and jeans she wouldn't open the door. She'd keep out the outside for as long as possible. A tinny clink sounded, the school bell. Oh God. The horrid schoolmaster was pulling the rope to summon the children. It was that late. What had happened to the alarm?

Laura scraped her hair back and tied a scarf tightly around it. She tucked in her shirt, draped a sweater over her shoulders and pulled open the door.

The surprise! All about her, overhead, behind her, the sky was a clear sapphire blue. Not a cloud. The sunlight made even the rubble and the pile of sticks beautiful, sharpened every crinkly bit of bark and shapeless wadge of mortar to a dazzling clarity. The changed world stung her eyes. Laura ran.

At headquarters she found Jenny in the front room sipping coffee, surrounded by bundles of dirty clothes. Laura stopped in the doorway and apologised; Annabel hadn't woken her, she moaned, and something had happened to her alarm clock. Was Bill furious? She must rush to the hill now and apologise to Christopher.

Jenny grumbled that she hadn't slept. The pithos on the roof rolled all night. She lay wide awake waiting for it to roll off but it never did. Bill was down at the pot shed finding work for everyone, she added, and she had the laundry to take to the fat woman in the ruined house by the sea, which she supposed she'd do when she'd finished her coffee.

'Aren't they working? What's happened?'

Jenny peeped at Laura over the edge of her mug and rubbed her chin down the warm side. 'What happened last night?' She was touched by Laura's consternation. She wasn't at all shocked by Laura's oversleeping, nor disapproving although she might have been if Laura wasn't so flustered by her mistake.

'Last night?' Laura blushed. She felt her face burn. Jenny smiled.

'You haven't missed anything,' Jenny put in kindly. 'They were nearly blown off the hill this morning and gave up. But Bill wants to keep everyone at it. I listened to him all night rehearsing what he'd tell everyone to do if they couldn't dig. "Stamp labels, wash sherds, finish sections, study finds." That's how he went on. I don't think he slept much. When I shouted at him, he complained that I'd woken him up. Make yourself some coffee.'

'Are you angry with me?'

Jenny kicked at the biggest bundle, which was tied up in a striped shirt Laura recognised as Bill's. She had been furious with Laura last night. But seeing her now in her pink T-shirt she wasn't angry any more. She was after all becoming more tolerant. She even managed another smile. 'Not with you. Only with all this washing!'

The two women carried the bulging bundles out into the brilliant day. The wind pushed them back. Jenny's dress was blown up to her waist; her feeble efforts to keep it down amused the men in the cafés. Laura was better off in jeans. They climbed over the toppled masonry and charred balustrade of the destroyed house to find the fat woman. Jenny called. Slowly the woman emerged from a small room still intact at the back. Hugely obese, wearing a brown sack with no sleeves, the woman let out a big belch before she ordered Jenny and Laura to come into her 'little house'. The flab on her arm jiggled as she beckoned to them to follow. Laura hesitated. Jenny whispered that the woman's family had been one of the richest in the village: her father was the mayor shot by the Germans. Their house was burned to the ground. She was the only survivor. The Germans had massacred all the men and animals after the Cretans ambushed seventy soldiers. Jenny stepped ahead with lips pressed together. Laura followed, made bold by Jenny's determination.

Inside, the fat woman set out two café chairs on the dirt floor. They were to have a sweet, and two spoonfuls of a white vanilla paste were handed to them. Laura gagged; the smell was sickly. Jenny licked at her portion as she stared at the stone wall in front of her. The fat woman described her favourite sweet tastes, asking what theirs were. She settled herself on her bed, opposite their chairs, and watched them enjoy her treat. Over the woman's head was a plastic baby with red fingernails and toenails nailed to the wall, and propped on the table was an ikon of the Transfiguration, the gold stars in a blue sky and the white cloud surrounding Christ very bright. When the woman saw that Laura was looking at the ikon, she grunted to her feet and pressed it to her lips. Would Laura like to kiss it, she asked. She sensed her embarrassment as Laura's chin pulled back. The ikon was quickly returned to its place and she sank back down on to the bed.

Laura was amazed when a thin squeal like a baby's came out of all that flab. 'How much will you pay me?' Her fat arms lay inert on her huge thighs.

'I will pay you,' Jenny assured her. Her Greek impressed Laura, but Jenny was terrified the Greek might fail at a crucial moment if she mentioned a figure. So she'd decided to let Bill bargain later when the washing was done.

'You've brought a lot of washing,' the woman pointed out.

Jenny admitted it; she shrugged and nodded her head once.

'It will take me two days. Two days at least. Maybe three.'

Jenny wished for Bill. Shouldn't she at this moment be warning the woman not to take too long, since they were all old clothes? Or was there something else she should say? Should she be sympathising with the woman? It was a horrid job. But Jenny feared that sympathy might put up the price.

The close, sweet smell in the room enveloped them in a sickly intimacy with the fat woman's farts and belches. It was finally too much, and Laura stood up and set down her spoon in front of the Transfiguration. Jenny swallowed the rest of hers and handed back her spoon. The woman followed them out into the ruined yard; the sea below, a steely platinum, was merely ruffled by the north wind. The women blinked in the light after the dark room, the wind butting and lashing at them where they stood beside the bundles of clothes.

'Buy me *sapouni*,' the fat woman squealed suddenly.

Jenny stared at the clothes, unable to think what the woman meant. Her eye ran up the twisted blue stripe of Bill's shirt. What was *sapouni*? Laura was out of earshot. Jenny turned to the woman and nodded. When the woman saw Jenny's perplexity she laughed. She pressed her arms to her belly and squealed that she'd buy the soap, she would need a lot. And then Jenny understood what the woman meant.

It was true. The mountains were like huge shadows chiselled out of the sky. Laura was so keen to take a walk that Jenny put off sweeping the headquarters' floor. Laura insisted she should get away. The fat woman was the limit. Jenny didn't feel, as Laura did, that they'd undergone something awful in the ruined house, but she agreed that the day was extraordinary.

Up the valley they went, walking fast. Past waving branches studded in pomegranates which were a dazzling red against the sky they walked on until they reached a steep path that led up to the old village fortified on all sides by crags of bare rock. Only when the people no longer feared pirates could they risk settling by the sea, Jenny instructed Laura breathlessly. Eventually they reached the high ruin of dry stone walls. There they were startled by a fig tree that filled a deep well with its white branches. Laura was tired and sat down on a wall to wonder at the tree.

'Are you tired?'

'Yes.' Laura felt beaten by the wind, and her eyes were seared by the light, which the wind intensified by brandishing shadows. Everything moved, blazed and floated, the light hard like the wall she sat on. She longed suddenly for the protection of a cave. 'It's too much.' She rubbed her eyes.

Jenny wasn't tired. She was happy, and incensed at Laura. The wind excited her; the light made her want to run and run, not feeling the thorns scratch her bare legs as she climbed higher to where she could see stretching on and on the mountains and valleys and the flat sea. In ecstasy as she ran her eyes across the arc of blue sky, it was hard to tuck her skirt under and sit down beside Laura. 'You're just tired, that's all,' she conceded, kicking at a stone.

'I'm sorry, darling, if I'm stopping you.' Laura laughed, running her hands down the sides of her face. 'I rather overdid it last night, I suppose.'

'Probably.'

With the tips of her fingers pressed to her chin, Laura shut her eyes. Jenny stared at those shut eyes, shocked by such passive exposure, such immodesty in the beautiful face of her friend. Quickly she looked away and stared uneasily at the bare branches of the fig tree. They were like an octopus the way they dipped into the well – a tangle of white arms in the light.

The two women slowly retraced their steps. In silence they picked their way through a sweet thyme that smelled of apple; small rocks skittered down as they moved their feet. They were both so preoccupied that when they came upon a freshly painted house half way down this desolate wilderness of ruins, it did not spring out at them until they were already quite near. They backed off, frightened by its inhabited look. Onions hung on one side of the arched doorway and on the other side a frying pan, which was so black against the whitewashed wall that it looked portentous. Up above, on top of a pole, flapped the black carcass of a crow. A dog barked. From behind the pole came the dull chortling of hens in a shed with a reed roof. In front, under the onions, sat a green plastic bowl and a mattress. They couldn't see the dog, but a short man with a thick, bare neck and bare arms came out of the door. He held his hand up to his eyes to scan below. When he caught sight of Laura and Jenny, he turned and ran, disappearing round the back.

'Why'd he do that?' Jenny broke their silence. 'Could he tell who we were?' They were too far away to recognise him, but the bare arms and neck squeezing out of a tight T-shirt looked familiar.

Laura was too lost in thought to answer. She'd glanced at the disappearing figure but wasn't surprised he wished to avoid them. They were intruders.

Only when they reached the road did Laura speak, the shiny leaves of the lemon trees and dark trunks of the olives at last protecting her from the wind. A black figure, who was stooping to pick beans, heard them and waved.

Jenny nudged a rock out of her way. 'What happened last night after you'd won?'

'Last night?'

'You were still playing when I left. Bill told me you won.'

'I didn't see you go.'

Jenny peered back at Laura who had dropped a little behind. 'I don't suppose you did!' She waited for Laura to catch up.

'What did Bill tell you?' Laura squinted into her friend's face. 'Did he tell you about a new workman called Michaelis? When Vasilis showed up with him, Bill was snoring with his mouth wide open, so I don't suppose he remembers much.'

Bill remembered a great deal. When Jenny asked him this morning what happened, he said he'd left Laura drunk, sprawled between Andonis and the new chap who had his eye on her. With a dirty snort Bill leaned back through the door to add that Laura wasn't 'averse to the humping male'.

Jenny kept her head down and murmured that Bill did the same thing at home. She no longer wanted Laura to pay for losing her the glory of the morning. Her surprise at finding that inhabited house near the old village made her forget her disappointment about the walk. Her suspicions about the bare-armed man who ran away from them excited her now.

But she was also anxious to know what Laura had done. She was even slightly afraid, like a child in a reptile house, in case what she heard disgusted her. She knew how intolerant she could be, and she knew her curiosity was not the curiosity of a friend. Jenny bent over to break off a sprig of the apple-smelling thyme which scented the stiller air under the trees.

'Doesn't it infuriate you? I've been there when he drops off after dinner. Not too flattering to the guests, I can tell you,' Laura retorted.

'Bill mentioned a new workman,' Jenny hazarded, unable to stop herself.

But Laura dodged the question. They both looked up. Above them a black cloud of birds flew over; the chattering deafened them as they dived into the trees. Then away up, up, they were off until they were a pale shape of tiny specks undulating with the current of blue air. Were they swallows? Laura was aroused. Jenny said they were; she could tell by their long tails. They were on their way to Africa. 'Lucky things!' exclaimed Laura. Jenny smiled. Their tennis shoes had just reached the cement of the village street.

'Laura!' A Greek pronounced Laura's name.

Jenny glanced behind. A man in a white shirt with red hair was waving to them; he was with Vasilis who was in the inevitable tight black T-shirt. 'It's Vasilis.' Jenny mumbled as she gave them a perfunctory wave and again faced the way ahead. But Laura had stopped. 'Why do they want *you*?' Jenny asked.

The men were catching up. Laura had shoved her hands into her pockets; she'd turned oddly shy, waving a leg idly at some fallen leaves. Jenny waited with her, puzzled. A goat a few feet from them was feasting on leaves blown off the mulberry trees which had been denuded in the storm.

Laura wasn't quite looking at the redhead but her body had tensed, her leg still as the two men stood with them in the middle of the street.

The redhead touched her arm. 'Where have you been? A walk?'

Jenny answered that they'd been to the old village.

'Why?' Vasilis blurted out, angry. He shook his small fist. Jenny looked at him suspiciously. Vasilis was grilling them. Had they seen anyone? Wasn't it a strange thing to do in such a bad wind? 'If you cannot work in such weather, why walk?' he shouted.

The redhead translated. Laura laughed. She was taking no notice of

Vasilis, riveted to the redhead. 'But it's a marvellous day,' she argued. The redhead laughed. 'Marvellous? Why marvellous?' he asked and he and Laura fell into a paroxysm of private mirth which excluded Jenny, while Vasilis fidgeted and glowered.

Jenny wanted to ask Vasilis who lived in the white house. Did she know him? She was taunted by the familiarity of that back view. If she knew who it was she'd know if he was running away from her. But she was frightened by Vasilis's anger, and thrown by Laura and the redhead giggling. She decided to pretend to Vasilis that they'd seen nothing strange. If that figure were Vasilis, which was what she now suspected, that would be why he had grilled them – afraid he'd been seen. Then much better that she should not ask. It would embarrass Vasilis and make her even more uncomfortable.

And who was this redhead? Was he the new chap Bill mentioned? Oh no . . . Jenny recoiled from the idea. She was desperate now to be off.

Vasilis laid his hand on the redhead's shoulder and ended the joke with Laura. He ordered the redhead to come along. They went on, apparently rejecting acquaintance with the two women with whom they'd just been conversing. The redhead was in Vasilis's power.

'What funny legs he has,' Laura remarked. She gave Jenny a conspiratorial look which was loudly out of tune with Jenny's mood. True, Vasilis's short back with his thick arms bursting out of his sleeves, and the redhead a head taller but with legs as short swathed in trousers too big for him, made an anomalous couple. Both men were bow-legged and pushed their feet forward like two old men.

'Who?' asked Jenny, at a loss to understand who had legs that Laura thought remarkable.

'Michaelis.'

Bill had looked everywhere for Jenny. He came out of headquarters just as Laura and Jenny, who were eager for a cup of coffee, reached the door. He lit into them. It was nearly lunchtime; the fat woman had been after him for rubber gloves and more soap; Mary Elizabeth had received a letter from her sister that her mother was dying. Where had she been? He'd even been up to their room in case Jenny had gone back to bed.

'What's she dying of?' Jenny asked.

'The fact is you should have been here,' he shouted back.

Laura told him to calm down.

'And you,' he glared at Laura, 'should be helping in the pot shed. Where have *you* been?'

Jenny and Laura hung their heads. Bill marched off down the street. Laura gave up on the cup of coffee and followed him.

In the yard outside the pot shed Ellen, Susan, Edward and Annabel were strewing sherds on trestle tables. The wind blew dust into their eyes. Empty cement bags cavorted between the tables, suddenly wrapping themselves around their legs. But they kept at it with heads down, picking out rims and bottoms and handles to reassemble the broken pottery. Susan was on her tomb finds, the others on their own finds from the Kallithea site. Martyrs to their work, they suffered discomfort with grim faces. Not a smile, not a word, not even more than a quick glance acknowledged Laura's existence. Laura wondered if the night before was on their minds, she the tart of the dig. Laura chuckled to herself (which Annabel heard) as she pushed upon the door to join the others inside.

'So, what do you think?' Bill was talking to Christopher. Adam stamped labels, hunched over a folding table in the corner. Christopher looked up and smiled; he offered his stool to Laura. He and Bill had been poring over a contour drawing of the cemetery hill pinned to Adam's drawing board. At the other drafting table Jack was drawing a deep blue, black and white bowl. Christopher saw Laura's eyes fix on the bowl. He told her to have a look. He ignored Bill's question, to follow Laura over. Laura picked it up; Christopher's hand jerked out to protect it from her. Unnerved, Laura carefully set it back and Christopher apologised.

'But isn't it wonderful?' he bragged. 'I'm thrilled, I really am!'

Laura moved to Jack's other side to see the bowl better.

'An Egyptian import of the Eighteenth Dynasty. Faience. It must be unique in Crete. Don't you think, Bill?' Christopher called. 'Can you think of any other like it?'

Laura hadn't noticed Mary Elizabeth in the shadows but now she broke in to disagree. Bill agreed with her that the bowl wasn't unique. They recalled another that had been found at Mavro Spilio.

Christopher enjoyed the sight of Laura giving the bowl her close attention. Jack informed Laura that it was about four thousand years old. The pure brilliance of the colours amazed her more than the decoration. 'It's a deep cobalt blue still,' she exclaimed.

‘Faience is the earliest thing like glass that we know.’ Christopher explained to Laura in detail the technique which was perfected, if not invented, by the Egyptians. Jack had stopped drawing. There was a hush. Even Mary Elizabeth didn’t interrupt. Christopher was justifying in an elliptical way his opinion that the decoration was unique. He addressed his words solely to Laura.

‘Now that you’ve had your say,’ Bill barked at him, ‘you had better come here and tell me where you’re digging next. I need Adam for my wall, so I hope you won’t need him to lay out trenches all next week.’

What if that whole hill were tombs jammed with Minoan valuables? Would Christopher be pleased? Laura could see the tombs were beginning to interest him. The bowl thrilled him. Perhaps there were more of them to be found at the bottom of Susan’s tomb. Laura joined Susan outside to help her sort, and Susan showed her what was a base and what a rim. ‘That bowl’s beautiful.’

‘I found it,’ Susan boasted. ‘I saw that the pieces joined from different bags and when I realised what I was finding I looked hard for the rest.’

‘Did you stick them together?’

‘Jack did.’

‘What happens if we find gold?’

‘And we’re going to,’ Susan whispered into Laura’s ear, just as a boy screamed ‘*Kyrie Klystopha*’ through the wire fence, holding up a Gordon’s gin bottle. Laura left Susan’s table to let him in; she also pushed open the pot shed door. Inside, the boy spoke to Christopher and handed him the bottle. He grinned with big gaps in his teeth and looked around him. ‘You are all archaeologists?’ he asked. ‘Have you found gold?’ Christopher shook his head and thanked him for the bottle.

‘From Vasilis. He made the raki the day before yesterday,’ the boy shouted. As he pushed back past Laura he stared at her for a good five seconds before he ran off across the yard and up the street. Everyone noticed. Laura pulled the door shut and returned to Susan’s table.

Laura separated rims and bases, putting bodies and handles in the middle and the coarse sherds in a separate pile. That little boy’s look hurt. What had he heard? Was Michaelis boasting? Had Vasilis described her evening with Christopher? Michaelis would pay! She’d worm into his gut like a monstrous Aphrodite. She’d make him crazy about her. Then see what they had to say at the café!

It had been Vasilis's idea to visit the still after Michaelis had beaten Andonis. Andonis refused to come. Vasilis led them to a lean-to near the police station where a tank sat on a fire, with rags wound around a spout which stuck out from the top with a rubber hose jammed on to it. The rags steamed. The hose, like a snake, was coiled in a drum of water, the tip of it drooling warm raki. An old woman handed Laura a pink teacup full, greeting Michaelis and Vasilis with raucous jokes Laura couldn't understand. Besides her there were two strong looking men in gumboots, two mothers in bright cardigans (one with a nearly newborn baby wrapped in a blanket) and several children. The women didn't drink, but all the men held teacups; their shadows on the low roof loomed in the lamp light.

The warm raki was strangely sweet. And lethal. Laura remembered how lethal from the morning after the *paneyyri* at Knossos when she had never felt so poisoned by alcohol. She wouldn't drink it down as Vasilis urged. She sipped, entranced by the swashbuckling efficiency of the two men in gumboots who sloshed water on the rags, uncoiled and coiled the rubber hose in fresh water, took the top off the boiling dregs to add more and pressed it back with bare hands. It was their second night. Michaelis showed them off to Laura as heroes. Talk dropped; comments from the old woman broke the comfortable silence. Laura stretched out her bare legs to the fire, settling into the chair one of the women brought out for her. Michaelis asked for something from the old woman. Laura shut her eyes. A little later Michaelis pressed her hand. 'Eat. Good,' he told her; she opened her eyes and saw him squatting at her feet with a stone and a bag of walnuts. He was hammering as if a queue of people were waiting on him. When he looked up and noticed she'd eaten what he gave her, he slammed the stone down to squash another nut. He pressed her to keep eating. It went well with the raki, he said. He refilled her teacup. It was true; the oily nut and sweetish liquid warmed her wonderfully. Laura walked over to the old woman and offered her a handful which the woman picked out of Laura's hand. Michaelis began to sing. The women grinned; Laura smiled at the women's pleasure. Vasilis left. The song had a chorus in which the men in gumboots joined. Laura began to hum with Michaelis, picking up the tune.

It hadn't occurred to Laura when she walked off with Vasilis and Michaelis that she would let Michaelis make love to her. She had noticed how the corners of his mouth pointed down, giving the lower part of his face a fine arrogance which she liked. Michaelis's brown eyes, under bushy red eyebrows, were lovely – they'd noticed her as soon as he came through the headquarters' door. Those serious eyes had excited her. But it wasn't until he sang and she caught the rhythm of the song with the taste of nuts on her tongue that the strangeness of the evening and his thin thighs caught her up. She forgot herself. Had Vasilis noticed the change in her? Is that why he left? Was it part of a plan? The low opinion he must have of her was frightening. Laura shied away from considering if Vasilis could have instigated what happened when Michaelis walked her back to her room. His hand rested lightly on her hip. She tilted her head so that it just touched his shoulder. A gust of wind blew them apart and whipped up her dress. He bent over to help her straighten it and when his hands curved round her thighs she laughed. Only his hand was steady; her hair, her dress, his shirt were in a frenzy.

If the wind hadn't ripped at them in the middle of the dark street, she wouldn't have been so overwhelmed by his steady hand. They walked out into the fruit groves where it was pitch black, the moon covered in cloud, the dark branches protecting them from the hectic sky. And on the ground there he undressed her. He took every stitch of clothing off her and, resisting her fairly desperate eagerness for him to come into her, he laid his head between her breasts and moved his hand down her body with amazing skill. Later, warm from his body sprawled across hers, she dressed, unfortunately catching her dress on a twig. She could barely see his face, it was so dark, and it wasn't surprising that she tore her dress.

CHAPTER 14

'Did he have a gun?' Jenny asked. She waved her knife at Bill to catch his attention. But Bill was laughing at something Annabel had just said at the other end of the table. 'Did he?' screamed Jenny. She recognised that flush on Bill's face. His babble and the outbursts, which were out of key whether indignant or enthusiastic, meant he was drunk. When he and Mary Elizabeth returned late she knew they hadn't just been telephoning North Carolina to find out about Mary Elizabeth's dying mother.

'Yes, he did,' Mary Elizabeth answered. Jenny could have slapped her face.

'Then it's the same one who was after me,' she retorted, slurping her next spoonful of soup. She raised her head. 'I thought he'd kill me!'

'All policemen in Greece carry guns,' put in Jack. How typical, Jack was thinking, of a hysterical woman like Jenny to think a policeman's gun was pointed at her.

'And,' Mary Elizabeth added, backing up what Jenny had just said as she glanced at Bill, 'he took out his gun and pointed it at Bill when Bill wouldn't move the car.'

Everyone sat still in amazement. Even Annabel gawped at Bill, who ducked into his soup bowl like a shy boy. Only Jack showed no surprise although he'd just criticised Jenny for falsely imagining such a thing. He thought it likely enough that a policeman would want to shoot Bill. His autocratic manner was infuriating. 'Tell us what happened,' Jack suggested with a smile.

'Did you report him?' asked Jenny. 'Did he shoot?' asked Susan. 'I suppose you moved the car then,' Edward assumed.

'He shouldn't be allowed,' Jenny shrieked. 'He terrifies those poor people in Ierapetra.'

'They sure were scared in the OTE,' agreed Mary Elizabeth. 'Weren't they?' She nudged Bill, which stopped Jenny from saying more.

'It's a government after all of little Hitlers. What's exceptional about that policeman?' Jack asked Bill.

'All policemen are an outrage to the dignity of a human being. Particularly traffic wardens! The lot should be strung up and shot,' Annabel blurted out.

'Is there anyone else you'd like killed?' Laura called down the table.

'Tax collectors,' piped up Edward.

'Social workers,' Susan put in. She surprised everyone. Ellen was shocked, and said so.

Jenny sat hunched at her place. Bill chomped at huge bites of bread apparently without a care in the world. The gun-happy policeman hadn't frightened him. How fatuously English. Or just drunk. When

Jenny had told Bill last night about the man who fled from her near the old village, she was called a silly goose for suspecting it was Vasilis. There was nothing suspicious about a freshly painted house. The man could have been anyone. Bill was so dismissive that she didn't go on to mention the strange meeting later with Vasilis and Michaelis. She would not suffer another put-down about what Laura was obviously up to with Michaelis. If Bill thought so little of intuitions, Jenny would be loyal to Laura and keep her out of it. Jenny smirked into her hand. If she'd begun with Laura, Bill would have listened. He loved to discuss Laura's scrapes with men. He found them predictable and amusing.

Jenny was glued to her seat as she recalled her terror in Ierapetra shopping with Margaret. The policeman hadn't exactly pointed his gun at her but he'd had one, and the way he raged at her for trying to park was as violent as if he had meant to shoot her. There was no reason for it. Why did Bill need to belittle her fear? Was he in the throes of impressing that pretty American girl?

Until then, Adam hadn't said a word, but worry and indignation constricted his eyes to slits. He was intent on every word. He sat opposite Jenny, who eventually noticed that he'd pushed his piece of cheese away to listen hard. He was about to explode. Increasingly aware and fascinated, Jenny braced herself.

As each new tidbit in the story fitted together in Adam's mind, the picture of what he detested most in the world grew, and excited his sympathy with Jenny's fear. A rush of affection for her pushed him into the argument.

Adam blurted out that Bill shouldn't have given in to an 'armed bully', an 'evil beast of a man', a 'torturer'; Bill had let himself be 'victimised and humiliated'. Adam hammered on the table. 'If you're English, you stand up to that sort of person. These poor Greeks have the Colonels. They can't. But we can,' he shouted.

'Don't be an ass.' Jack started to laugh. 'That policeman could see a mile off that Bill was another damn *milord*. Why not have a shot at him?'

Laura was collecting plates. She pointed Jack's at his nose. 'Because it's not a very nice thing to do,' she hissed.

'He wasn't trying to be nice.'

'He could have asked Bill to bugger off in a nicer way.'

'That's up to the policeman.'

Adam's face turned puce; the veins in his neck bulged, in the dull light: 'What do you mean?'

'If he wants to point a gun at Bill, it's his gun.'

Uproar broke out. Others were on their feet now. Laura tapped Jack on the head with his plate. Jack grabbed her arm and the plate broke on the floor. Now Jack too was shouting. The noise in the small room had reached a pitch of crisis that was ludicrous, Jack on a chair, Adam with both fists raised, Edward jabbing the air with his fork. Susan threw an apple at Jack which missed.

Bill cut himself a large chunk of cheese and asked Laura to hand him an apple. Mary Elizabeth thanked Bill for the trip into Ierapetra and slipped away. She left the door slightly open. The German stuck his head through. He and his friend had heard the noise as they were passing the archaeologists' headquarters on their walk up the street. 'Good evening!' the German hollered. The big blond head came into the room. His glistening face stunned everyone. There was silence. He swung his arms and laughed. Could he and his friend join the party?

CHAPTER 15

The morning after the storm, after a bad night worrying where to put down new trenches (his discussion with Bill as they examined Adam's contour drawing in the pot shed had made him uneasy), Christopher found Susan at headquarters drumming on the table with her fingers. She poured him a cup of tea and shoved the jar of marmalade at him. She'd already eaten. The wind had stopped, as suddenly as a balloon withers when it loses air, as Christopher walked up from his room.

'We have a bit further to go before we reach bedrock,' Susan chivvied him. She stood up with her plate. 'I expect to find the top half of my dagger today. I'm off.' She banged the door behind her.

Christopher carried his plate and cup to the kitchen and stepped out into the cool dawn. From the café next door a commotion of men's voices carried out into the grey street. The air was uncannily still. They would finish clearing Susan's tomb, lay out the trenches, take off the surface of the new trench on the west side, and complete the final section of the robbed tomb and Susan's tomb so that she could supervise a new trench. All night Christopher had tried to decide which side of the hill would have attracted the Minoan dead, the side nearest their village or the side facing the mountains? Not the sea side, which would not have promised security, so he could dismiss the south slopes.

When he reached the orange groves, he could see Laura's blue anorak ahead, moving slowly down the middle of the road. Christopher liked it when Laura scraped the hair back from her face and tied it in a scarf as now, her wide face, with its high cheekbones, beautiful. He looked down at her, pleased to find her alone on her way to work.

The workmen were smoking by the olive tree when he and Laura made it to the top of the hill. The rim of the sun had just risen out of the sea, bathing their faces in orange light.

'Come on,' Christopher ordered. He broke the spell of sunrise, stabbed by the worry of time missed because of the storm. 'Let's get on!'

Christopher waited impatiently for Adam, who had gone with Bill first to plan his wall. He paced the hill, scrutinising the ground for ant holes. He found several ant holes but on no alignment; they did not connect in any visible way. On the other hand, if the new trenches were to be on a proper alignment, which would be the orderly way to lay them out, they might turn out to be an archaeological fantasy and miss the tombs altogether. This was one of the reasons he disliked digging tombs. It was difficult to avoid the greed of the treasure hunt – which he found distasteful – and keep to a scientific approach, particularly when the science of aligned trenches could prevent their finding what they were there to find.

Christopher studied a stone for a minute to calm himself. He was too anxious. He mustn't be a dithering fool. He pulled back his shoulders and took deep breaths of the morning air, shifting his gaze to the purple mountains. The magical island of Crete with its beautiful mountains and caves and secrets was the conundrum. Never mind alignments. He'd put trenches down where there were ant holes.

Andonis shovelled for the new red-headed chap. They were moving earth at a fantastic speed, speaking hardly at all, which surprised Christopher when he walked over to see. They were competing with each other like sullen schoolboys.

'You're finding nothing?'

'Nothing. It doesn't make sense.' Susan stood behind Manolis with her hands on her hips. 'The surface wasn't blown about enough,' she snarled. She looked like an angry old woman.

Christopher asked Manolis what he thought. The boy held up a fistful of earth to show how loose it was. Susan climbed out with another bucket of earth to be sieved. She sighed when she set it down beside Laura. Christopher pinched the earth in Laura's sieve, thinking that Susan had expected too much of this last level of her tomb. The wind blew in dust which gave the surface the smooth look she distrusted. It was the wind also that made the dirt feel like sand.

What he would resist at all costs, Christopher told Susan after watching a while longer to see how Manolis got on, was any suggestion that it had been robbed.

'That's what's happened,' Susan retorted.

'No, that is *not* what's happened.'

Manolis was caught by the impatience in Christopher's voice. Susan looked obdurate, her head right down.

Christopher left them. Adam at last had arrived, his red face blotched white from heat. Christopher took Adam's drawing board so that he could tilt the water *stamna* and crouch under it. 'Where've you been?' Christopher was irritated by the delay, all the more because of the possibility that his new trenches would miss the tombs. It was like not knowing if you were on the right road – you wanted to get to the end so that you could try another if it were wrong. 'You're late.'

'Bill's found a paved floor at the base of his wall. He was at my door at 4.30 this morning!'

'Not about the paved floor.'

'A new trench to see how far the wall extends. Just now they reached paving.' Christopher forgot he was holding Adam's drawing board. 'Actually Edward and Annabel have also found a tablet and some bronze thing.'

Christopher grilled Adam on the tablet, which Adam thought was Linear A and whole; Badger had missed it but the local boy noticed it when he dumped his barrowload. 'And what does Bill think?' Christopher harangued Adam, who was still cooling himself under the spout of the *stamna*.

Adam rose slowly to his feet. 'He's taken up with his pavement.'

Christopher handed back Adam's board. 'It's amazing that a tablet should turn up on the edge of the site. Bill must think the site continues to another important building. How fascinating!' Christopher blurted out.

Adam jumped. Christopher laughed. Bill had been so frustrated at losing a day's work.

More slowly than if Adam hadn't brought such exciting news, Christopher and Adam walked from ant hole to ant hole, looking behind them to see how they'd come and how the hill and the holes connected. Two more trenches were decided on. There would now be three trenches for Christopher and Susan and Edward, if he came, to manage. The new workman Michaelis was doing well; Christopher would keep him and Andonis for the heavy work of moving earth, and Manolis would pick. One good pickman should do. Bill would need the others now more than ever.

Over second breakfast at eleven Adam described how Bill had raced about, shouting at everyone and furious at Badger. Christopher corrected Adam since he knew that Bill would not have blamed Badger. Such things often happened. It meant a good reward for the barrow boy. No, not furious, Adam conceded, but upset. Bill had put Annabel to sieving.

'Like me.' Laura raised her arm. 'Sieving's for the ignorant. But never you mind, the meek shall inherit the earth.'

'The sieve is hardly the earth,' retorted Adam. 'Have you found anything?'

Laura looked at Susan who scowled and shook her head.

Christopher stayed to study his notes after the others went back to work. He asked Andonis to hold the ranging rods for Adam since Michaelis, a fast worker, seemed able to carry on on his own. Intelligent also, Christopher thought, judging by the way he picked evenly, thinking in terms of the whole square. He did not burrow in one corner like an animal. Impressed, and sad that he wasn't with Bill at that moment, Christopher watched Andonis stand as still as a statue as he held the ranging rod while Adam scampered about measuring and banging in nails. He wondered if he'd done something to a tendon and kneaded his aching calf, remembering his hypochondriacal mother who had once told him – over breakfast when he was home from school one Easter and had unwisely complained to her that he'd probably not be put in the tennis team – that if he didn't watch out he'd always be a loser. Like his father, she said. His father never risked competing; he'd let people walk over him and have everything. She was only a loser because of her health. To hear her use the slang term 'loser' in itself had shocked him. Perhaps she'd been reading a novel about 'losers'. It had been a rainy spring morning, and he'd had no way to throw off this curse of hers; he could remember the blue china coffeepot and the triangles of toast in the silver rack sitting between them as she ridiculed his youth and promise and expectations. He should have kept quiet. He hadn't minded at all about the tennis team. Of course now he could smile about it because he could see she was right.

Christopher slapped the notebook against his leg and stood up. As he dusted himself off he decided to ask Laura if she'd like to change places with Annabel so that she could be where the action was. He started to walk over to her.

There was a shout. Someone called, 'Christopher.' A tall man in a suit waved from across the hill.

It was the Ephor. Christopher forgot about Laura.

The Ephor was a huge man with wide drooping shoulders. He was a scrupulous dresser 'in the English manner', his bent posture giving him the look of a harassed scholar. He was known to like foreigners and admire their work when it was good. If the Ephor himself weren't such a good scholar he might have been less irascible, since his job kept him from his work. Christopher had first met Koraes when Koraes was a student at the University of Thessaloniki, which was a long time ago.

'Have you heard?' Christopher shouted.

'What?'

'What they found this morning?'

'Who?' The Ephor was disconcerted by the rush of enthusiasm. It wasn't like Christopher. 'I've come to see you.'

The Ephor listened, but surprised Christopher when he said he hadn't the time to go and see. Christopher was to make sure the tablet reached him at the museum as soon as possible. That was all. He made Christopher uneasy. Why had he come?

They paced the hill together. The Ephor said little. He would look up and then down again, as if the sky could give him a reading on the undiscovered tombs. Christopher introduced Susan, giving her a hard look which she seemed to understand. She described effortlessly to the Ephor the objects they had found and their date, pointing out where they'd lain in the tomb chamber. Christopher was even a little surprised that she was supporting him in front of the Ephor. Laura gave the Ephor a big smile, holding out her hand so graciously that the Ephor smiled for the first time and asked if she'd worked on many excavations. When she described herself as a 'debutante' and a 'fellow traveller' he was delighted. Christopher backed away, pretending he had something to do, to give Laura more time. A mollified Ephor would be easier. What his business was Christopher didn't yet know, but he feared there was something up.

Adam and Andonis had worked out a system with the tapes. Adam shouted to Christopher to look out for red cartridge cases which could mark the measuring pins. It was midseason for shooting song birds and the ground was covered with empty cartridges. Christopher handed Adam what red ones he found and walked on to have a word with Michaelis. The Ephor caught him up.

'Is this all of your team then?' he asked.

It was Adam's last day, Christopher explained, but Susan would carry on. The Ephor could see how competent she was.

'And the workmen? You know them?'

Christopher assured the Ephor that he knew them well. Andonis he'd known for a very long time and Manolis, though young, was good. Christopher and the Ephor had moved nearer to Michaelis's trench and were now standing only a few yards from him.

'He's a Greek?' asked the Ephor, sounding incredulous.

'From Ierapetra.'

The Ephor asked Michaelis why he had red hair. Michaelis threw down his pick and pulled at his hair. He broke into a tirade about how people mistook him for a German, which was a lie and an insult and a sadness. Fascinated, the Ephor moved closer to discuss with Michaelis his genes; he grilled him on who else in his family had the same colour hair. Christopher had heard that the Ephor was an amateur genealogist, an expert on 'cousinage' for instance, but he'd never before witnessed the Ephor at work.

'So.' They were walking back to the Ephor's car. 'You have three workmen all of whom you say you are sure of. You know that redhead too?'

'He's good,' said Christopher. 'You see all he's done in a morning. And intelligent.'

'But you *know* him. That's what I'm asking you.'

'Yes, I know him. He's a cousin of the café owner Vasilis who's been a great friend. He has helped Bill in all sorts of ways; Bill bought the Kallithea site from him, or much of it.'

'From Vasilis. Not from the redhead!'

'No.'

'You see.' The Ephor stopped and stared for a moment at the ground. Christopher had that sick feeling he remembered from school when he stood in front of the headmaster.

'I will be sending you someone from the museum. To help, of course.'

Christopher said he'd tell Bill, and thanked the Ephor on Bill's behalf.

'But you're in charge here, no?' The Ephor's finger stabbed the air. 'This is your dig. You are the director.'

With a diffidence that usually pleased the Ephor, Christopher shrugged.

'*You* are, *you* are!' the Ephor shouted at him. Suddenly he was angry. Christopher couldn't think what he'd done. He apologised. The Ephor waved his apology away and thrust out his hand. With an abrupt goodbye he was off down the hill, stones skittering in the path of his well-polished shoes, the huge well-dressed figure an anomaly in the barren landscape.

Susan was waving. Christopher walked over slowly, trying to collect himself. He fumbled in his pocket for his flask and took a nip of brandy.

'I have something to show you,' Susan called. He put his flask back as she handed something to him. 'And we've reached bedrock so that's all there is. Or all that's left . . . anyway.' Her voice trailed off, snuffed out by a twinge of shame.

Christopher turned it over in his hand, the base of another faience vase, the pair. He grabbed the sieve from Laura and shook through the earth. He pushed it through quicker with his hand. He jumped down into the *dromos* and grabbed Manolis's pick to scour the sides and corners of the chamber, leaning as far in as he could.

'Is there more earth to sieve?' He darted for a bucket of earth on the other side of Laura. He emptied it on to the sieve and pushed it through. Laura and Susan watched in silence until he'd finished and straightened himself. He was scowling, hands on his hips, much as Susan had looked before the Ephor appeared.

Bill wouldn't believe it. Then he wouldn't believe Christopher had told Koraes about the Linear A tablet and the wall. He left his lunch untouched to fight disappointment and fury. 'And after such a day, this!' Bill wailed, in the whisper only Christopher would hear. Christopher was very sorry and decided not to let Bill know how much he, too, was upset. Bill spread out his hand. 'What did he mean?' Christopher shook his head. 'But everyone knows when you visit an excavation you go to the director who shows you round.'

Quietly Christopher agreed that it was not the way to behave.

'So what am I supposed to do now? Am I in disgrace?'

'Of course not.'

Bill stared as his jaws chomped on the meat, and Christopher pitied him. Bill minded so about his authority. And while in this turmoil Bill wouldn't be able to explain the Ephor's motives. But that would come. Christopher knew Bill well. Bill would take relief in exercising his cunning and would insist melodramatically on the truth of his dogma, plotting his conduct accordingly. The Ephor would be nailed, according to Bill's version, to save the dig. This Christopher dreaded. He wished Bill's hurt wouldn't transform itself into distrust. Probably the Ephor had had a bad night or his back ached. Those were the sort of interpretations Christopher preferred; they didn't burden the feelings to the extent that Bill's jumping to conclusions did.

Weighing on Christopher's mind more than the Ephor's visit was that lone base of another faience vase. Up on the hill he had been emphatic to Susan that it was an old break. It was impossible to tell what had happened to the rest. The important thing was that the dead man had had a pair. Laura laughed, hoping no one would bury her with the chipped lid of her Spode coffee pot. Which, thank goodness, had made even Susan smile. Christopher regretted that he'd revealed his disappointment when Susan handed him the base. He must now be on guard with her. He would not tell any of this to Bill since in his present mood Bill was dangerous.

Margaret had cooked a delicious lunch. She'd made a sauce for the pork chops which was light and sharp and delectable; the spinach purée and gratin potatoes went well with it and Christopher tucked in, convinced it was one of Margaret's best meals so far. Everyone enjoyed it, scraping their plates clean, the 'mmmm' sounds of gratified eaters the only talk until all the food was gone. Too tired to talk, and too hungry, the excitement of the day's finds held them all in thrall (except Bill and Christopher of course), a silent satisfaction drifting out through the open door to the couple of newly arrived hippies who chewed at pomegranates in the middle of the street until they were chased off by a woman with three goats, whacking the mulberry trees with a long stick.

There was still Margaret's apple tart. She'd outdone herself. But Christopher noticed his hand shook when he started to cut into it. He gulped down more wine and shut his eyes. A talk with Bill alone to help him see sense and not waste energy on anger. Need he? He wanted to see the staircase and pavement. Go together now after lunch? Should he suggest it?

Christopher suggested nothing. He pushed his plate away. Out in the street the afternoon sun blinded him, his eyes were so tired, and he rubbed the back of his neck, stooping like an old man. He felt like cardboard, stiff in the knees and shoulders and wrists. When he reached the pot shed he collapsed on his bed with his shoes on.

Much later when Christopher woke up, it was dark. The day was finished except for supper. And then more sleep. He sat dazed on the edge of his bed. Something. It came back, spreading through him like a drop of ugly colour muddying his fresh calm. He groped his way outside to the tap and splashed water on his face. Bill's rage. There was still that. He'd done nothing yet. Could he skip supper? Go back to bed now? Put off their talk until tomorrow? He drenched his trousers as he cupped his hands for more water, also wetting his shirt. He must face Bill. Now.

The relief when Bill came in late to supper flushed from drink and pleasure was huge. A wonderful shock. He was with the pretty American who, Christopher had already noticed, pleased him. When the two of them squeezed into their places with smiles on their faces Christopher was overjoyed. Well done, Bill! What a sensible way to get over the Ephor's insult. How typical of Bill to surprise him after he'd so dreaded his coming. Christopher resolved to keep clear of Bill now for the next few days so that a discussion wouldn't start him up again. For reasons of diplomacy and pride Bill would not discuss the Ephor with anyone else – of that Christopher was sure. So, if Bill didn't find Christopher to go over it with him alone, the Ephor's visit (like the incident with the mad policeman which Christopher saw as supper went on hadn't upset Bill) might be forgotten.

Annabel cornered Christopher as he was leaving to suggest she walk with him a little before she went to bed. Once outside she suggested they go to Kritsa for the weekend so that he could show her the museum at Ayios Nikolaos and the church of Kera, with its famous frescos, as well as Gournia and Lato. They were all places she was dying to see. Christopher saw his chance. He jumped at it. He was grateful to Annabel, even if it meant putting up with her flummery. She was a good sort despite that. They decided to catch the three o'clock bus after lunch tomorrow which was, thank goodness, Saturday.

CHAPTER 16

Bill pointed his torch at Laura's sandals. 'You'll twist your ankle in those. Jenny's been looking for you. She's gone now.' He sounded as sore as a tied-up dog.

'What's the gripe? I'm tired.' Bill stood in front of their 'houses' as Laura climbed up the rubble heap. She walked past and pushed open her door. As she recoiled from the dark room, she bumped into Bill, who picked out circles of unmade bed and crumpled clothes with his torch. 'Hold it there!' Laura tripped over a suitcase of Annabel's, then stepped on a shirt until her hand knocked against the smooth glass of the lamp.

'Want a drink?' Bill was careful not to step into the room. 'I have a bottle of whisky here, if you like.' Doing a typical Bill, he suddenly turned away and left Laura on her hands and knees in the dark. 'Come out and I'll pour you a drink,' he called. Laura heard the scrape of a chair on the pavement outside. She stumbled out with her lamp and asked him for a match.

'Neat?'

'A little water.'

Bill used the tap in the loo and handed Laura her drink. They faced the dark on chairs a yard apart. Bill beamed the torch on the chickens; they untucked their heads and clucked.

'Don't torture the poor things!'

Bill snapped off the torch, and they sipped whisky.

'Why didn't you go in with the others?' Laura asked eventually. The drink warmed her after her walk.

Bill sighed. He'd set Laura's lamp down on the far side of him so that the outstretched legs, the woolly head, the stubby nose and the protruding chin stood out. What a bullish profile! Here was the hardy Englishman, 'a man of conviction'. Laura smiled, a smile Bill couldn't see. 'I have a lot on my mind,' he replied.

'On the whole you're pleased, aren't you?'

'On the whole.'

Laura tittered. 'There is that trigger-happy policeman who prowls the wastelands of Ierapetra.'

'He's not a worry,' Bill snapped.

'I thought he sounded rather an ugly sort myself.'

'Jenny was sorry to miss you. She had to drive Adam in for more graph paper,' Bill grumbled.

'What's wrong?'

Bill pulled in his legs and cleared his throat so as not to say what was wrong.

'Don't say if you don't want to.'

'The Ephor. I've done something wrong,' Bill blurted out. 'He doesn't like my wall. It proves him wrong. His peace-loving Minoans who lived in unfortified villages are a dream, but he's so set on being right that he'll have my permit withdrawn rather than acknowledge his mistake.'

'Really?'

Bill ground his shoe in the dust, hunched over his knees. He set his glass down between his feet.

'Rather a nice man, I thought.' Laura pursed her lips to keep back a smile. 'Natty dresser.' She sipped her whisky. In a dull sort of way she was enjoying the two of them talking alone together. Bill was sounding so familiar. Anyone important like the Ephor threatened him. He was the same all those years ago in Oxford when he'd inveigh against a tutor or a college rule, even when he was kneading her bosom. A wrong could occur to Bill at any time.

Did Bill remember? He must, but without sentiment. Did she still feel anything? The loss of youth. Bill and she had been young. They were both virgins when he'd had his scout prepare a lunch in his sitting room at Magdalen their first term. Oxford was still new. It was a rainy Saturday. They'd met for coffee at the Cadena. They'd walked through the market and up the Turl, pretending they were tourists. They looked in at Jesus and Lincoln and Hertford to compare architecture, bending their heads right back to inspect cornices, pediments, traceries. It was fun. They'd transformed that grey Saturday into just an ordinary Saturday by pretending they weren't undergraduates. They visited the small science museum and the Sheldonian and climbed to the top of the university church tower. They'd reeled off what was what; he'd picked out Magdalen first of all, and of course she'd pointed out Lady Margaret Hall. They laughed to see how loyal they were. At the white tablecloth in his rooms they were waited on by the scout. Bill told him grandly that he did not want to be disturbed. When the scout had bowed himself out, they giggled and settled on the sofa for Bill to start awkwardly kissing her and stroking her hand while they discussed Henry James's short stories. On the floor they grappled with his zip and her blouse buttons, unequal to the occasion. She didn't have qualms about losing her virginity, a 'virtue' to which she'd never attached value. And Bill was desperate to lose his. They made it and afterwards were so proud that they rushed back into their clothes to walk in Christ Church meadow and discuss love.

Had she ever been in love with Bill? For a little while. That first term, once they'd deflowered each other, they met every day if possible and made love, which wasn't always easy. Bill worked hard and was ambitious, his heart set on getting a first. He'd eat in college and afterwards work in the evening in his rooms. They met in the afternoons when he could spare the time. She soon got bored. They met a few times in London during the vacation but things were cooling and by the end of the second term they only met occasionally at parties.

It was a surprise when her great friend Jenny began to go out with him in their last year. At the start Laura imagined she must know more than either of them about their friendship. She held the strings when her first lover and best friend began to go out.

She'd wanted Jenny to find someone. Jenny's appearance was so extraordinary after her hair turned grey their second year, with her green eyes and smooth dark face, that she'd attracted men easily. But Laura had guessed that Jenny could never let men further than a passionate kiss. Always boyfriends were short-lived and her loves, by the time she did fall in love, were unrequited. Often at Oxford she called herself ugly and lumpish. They'd sit up until late at night discussing either how she might make a success of a new boyfriend, or why she'd failed to.

The Saturday evening in their last term when Jenny came back late from punting all day with Bill, flushed from wine and kissing, and confessed to Laura they were in love, Laura wasn't thrilled but pleased. She thought she knew how Bill had behaved with Jenny, in the same impetuous, self-centred way as he behaved with her – which consoled her. To say she felt betrayed was going too far, but there was a sense of betrayal lurking in her less conscious self when Jenny and Bill's friendship turned into a romance. When she next met Bill and saw how Bill ogled Jenny and forgot to sip his tea as he sat cross-legged on her floor, Laura was shocked. It was a different Bill. Jenny had roused a solicitous and foolish side that Laura had never known. Eventually she could enjoy the thrill of Jenny's finding a man, but she had first to forget what she knew of Bill Courage from an earlier time.

In the semi-darkness, Bill's glum figure looked abandoned and misunderstood. He had always had that adolescent pitifulness about him. 'Why didn't you go in with Jenny and cheer yourself up, darling?' she asked, crossing her legs the other way to make herself more comfortable.

Bill scoured the pile of sticks with the torch, and then snapped it off.

'Weren't you cheerful last night after your whirl with Mary Elizabeth?'

Bill stood up. 'I should have kept my mouth shut. I'm a fool!' He picked up the whisky bottle. 'As if I haven't enough on my mind, Jenny called me a coward and threw my cufflink box at me!' He rubbed his eye.

Before Laura had found Bill with his torch and whisky bottle, she had sat behind the church trying not to feel sorry for herself as she listened to the sough of the wind in the pine trees. She had already walked barefoot on the sand at the edge of the sea, the breeze in her hair, before climbing up to her bench which faced the bulbous apse and pink dome of the small white building half way up the hill. The sky in the west had turned the same bizarre pink as the dome, with wild gashes of purple cloud disgracing it. She watched a frail moon, like a meagre shaving of light, waiting for the colour to go, and felt herself dramatically alone until two boys climbed the wall and babbled about her. How odd she looked, a lone, middle-aged English woman in jeans and sandals. It started her back to her room to lie down.

If Annabel had not been so loud at lunch about what she and Christopher were going to do and see together, Laura would have finished her chicken salad and not walked out on everyone. The two of them were together now in a car having a good time. What had come over Christopher? Spend a whole twenty-four hours alone with that trout? Laura was mortified. In fact everyone had left on the three-o'clock bus, Mary Elizabeth off to her handsome Epimelete, Jack and Susan to some cheap hotel in Ierapetra, Edward and Ellen off. Everyone except her, because she was Bill and Jenny's friend and too old. In her present mood she might post her letter to John; so far she'd stopped herself, knowing it would be a mistake.

But what was there to hold on to when even a priggish academic like Christopher let her down? What a hypocrite! She could have sworn that Annabel was the last person he'd want to go off with.

As she walked along the sea with water and sand squelching between her toes, Laura went over her relations with both John and Francis. With John there'd been sex, which brought her closer to him than anything else. But their lovemaking hadn't made her indispensable. He went back to his wife when she challenged him. And Francis, her husband, grew to hate her when she wanted him to love her as much as she loved him. He took to drink and finally came at her with a broken bottle. The mad hate in his face was horrible. She ran out into the street in pouring rain in her dressing gown. Neighbours saw. She ran to the end of the street across the traffic; a bus squealed to a stop. The newsagent was still open. Inside the wife in her sari with a red spangle between her eyebrows gaped until Laura explained. She took Laura's hand and led her into a little room at the back to make her a cup of tea and settle her on the sofa with a towel and some magazines.

Why did men fear love? The newsagent's wife understood although her English was poor. Laura spent the night on her sofa, and she lent Laura her coat the next day. Laura had been a fool to make Francis the centre of her life and allow him to give sense to the fleeting jumble of days: she'd put the double amount of finely ground continental roast in the filter paper every morning just as he liked it, and was pleased with herself because she'd pleased him. It gave him indigestion. She'd got it wrong. By the end Francis couldn't stand the sight of her.

Laura held up her foot but it was too dark to see the red of her sandal any more. 'Where do you think Mary Elizabeth went this weekend?' She squinted up at the black figure blocking her view of the night.

Adam wanted to have a drink first. He'd found graph paper while Jenny bought shampoo and a newspaper and waited for him under the tamarisk trees in the new part of town. Jenny had already started on a bottle of retsina. Adam ordered a *carafaki* of raki and dragged up a chair.

'Jack and Susan were terrified I'd force them to have dinner with us,' she remarked as offhandedly as she could, pouring out more wine.

Adam smiled.

'Am I that terrifying?' She hoped it sounded an idle question.

The waiter brought the raki and Adam filled the small glass from the carafe. He gulped it down, coughed and poured out more.

Jenny watched Adam surreptitiously. What would be safe to talk about that wouldn't bore him? She felt ungainly alone with him. She pushed back her chair a little and crosssed her legs the other way as she searched her lap for inspiration.

'How long have you and Laura known each other?' Adam was asking suddenly. Jenny looked up in dismay. Laura was far too personal to discuss with Adam. She drank her wine, told him more than twenty years and refilled her glass.

'We're very different,' she added politely.

'She's the most terrific woman I've ever met.'

'Yes, she is.' Jenny pressed her lips together.

'I wish she'd hit Jack really hard with the plate that night. He's a smug shit!'

Jenny giggled into her glass. Adam was staring at her, which made it difficult to relax.

'What's it like being married to Bill?' he asked now. 'Are you happy?'

What impertinence! Jenny mustn't answer such a question. Where on earth had Laura gone, leaving her alone on a Saturday night with this boy? In a panic Jenny shook her head. She even missed Bill. The thought of him made her sigh. She'd thrown the box of cufflinks because he'd pretended everything was normal. For Christopher and Annabel to go off together was normal. The fight at dinner, he and Mary Elizabeth, the policeman and his gun were all *normal*. The leather box hit him in the eye. He sank on to the bed holding his head. She stomped off. Half an hour later, as calmly as if nothing had happened, he waved goodbye. He'd come down to ask if she'd buy him a newspaper. His eye was still red.

'I hope the cufflinks weren't hurt. My father gave them to Bill when we were married,' Jenny mumbled.

'What are you talking about?'

Jenny flared up. 'I'm talking about the gold cufflinks my father gave him. You're drinking too much.'

Adam laughed. Jenny blushed.

'Don't worry,' said Adam. 'I'm not laughing at you. Bill can't be easy.'

'I'm not especially easy myself,' Jenny interrupted. 'Why can't Bill be easy?'

Adam called the waiter who with the second *carafaki* of raki brought more nuts and a plate of cucumber. Jenny relented over another bottle of retsina. The night began to melt. Jenny sensed someone a few yards away, but was not bothered enough to turn round and see who it was. It was Saturday evening after all. There were always lots of people about then with not much to do. She had heard the figure pass them and his footsteps stop and the scrunch of gravel die away, which left him close by. But since boredom cluttered the cafés, it was important not to mind inquisitiveness.

'I'm happy,' Adam announced.

Jenny stared at her glass.

'It's a terrific feeling.'

'Oh.'

'This week especially. Bill doesn't know it yet, but I understand his site. I see the views the Minoans designed around for one thing.' Adam's hands cupped pockets of darkness and impersonated walls and roofs. His arms suddenly stretched out to embrace a courtyard at the bottom of the staircase which hadn't yet been found. 'Bill's too cautious. He can't help it.'

Jenny watched Adam without listening, fascinated by his angry frown and the way his head waved with his words. Light from the café caught his eyes and the ring on his finger; wind in the feathery tamarisks waved shadows across him and the table. 'I envy you.'

'But it's your Laura really.'

Jenny pulled back her chin.

'I'm in love with her.'

'You can't be,' she snapped. She suggested that they look for a restaurant before it got too late. She stood up, clutching the shampoo and the newspaper. Adam sprawled back in his chair, his long arms dangling over the back like wet washing. He winked at her and pulled in his legs to reach into his pocket for his wallet. Out of the shadows behind, the scrunch of gravel started up, one step and another, right up to their table. Jenny peered around. The policeman stood a foot away, the same policeman who had made her move her car, the one who yesterday had waved his gun at Bill. He stood in front of them with his arms crossed. Had they done something wrong? Jenny urged Adam to hurry up. Adam asked her what the hurry was. She moved round to his chair and tapped him on the shin with her foot. 'Come on!' she whispered. Adam wouldn't move. His hand looked stuck in that pocket, no wallet retrieved. He leaned his other arm on the table with imbecile insouciance. Jenny opened up her handbag. 'I'll pay.' She kicked him again under the table.

The policeman continued to stare and it wasn't a bored stare. He was watching to see what they would do.

'Please, Adam!' Jenny waved money at him and looked for the waiter. They were the only people left. She squeezed the money into Adam's hand and ordered him to go and find the waiter to pay.

Adam's hand was still in his pocket. His head was bowed and his eyes fixed on the empty *carafaki* in front of him. Jenny realised that he was about to do something and was appalled. She was frightened. She leaned over to whisper into Adam's ear that the policeman had a gun. Adam threw back his head and laughed. Very slowly he rose to his feet. But, when Jenny thought they were off, Adam propped his arms on the table and broke into laughter. His shoulders, his back, all of him heaved.

His act was so artificial that Jenny was astonished when the policeman asked Adam what was funny.

'You! You're the funniest little puffed-up Hitler I've ever seen.' Then in Greek: 'I don't like you. I think you're stupid and mad and a disgrace to humanity.' He was still guffawing when he straightened up and handed Jenny back her money. He pulled out a 500-drachma note and slapped it down on the table; he saluted, clicked his heels and slipped his arm around Jenny's waist, walking her off with him as if they were sweethearts. Jenny tripped on a stone, but they didn't stop. Jenny was too scared to look back.

When they had finished their dinner and were sitting on the quay of the old port listening to the creak of the fishing boats and the slop of the water in the warm night, they realised they had been followed. By the street lamp Jenny saw the visored cap and the black holster. She clapped her hand over Adam's mouth before he could speak. Adam pulled her hand away and kissed her on the mouth. 'See what he makes of that.' He took Jenny by the hand and led her past the policeman to where they'd parked the car. Nothing more was said until they were inside and Jenny had started the engine.

'Sorry about that.' Adam had noticed how Jenny shrank back when his lips touched hers.

'You're drunk,' she retorted.

Adam giggled. 'Policemen hate lovers.'

Jenny almost hit the bumper of the car in front; her lights caught the waving heaps of tackle on the boats below. She inched past carts and motorbikes down a narrow dirt street, relieved when she made it to the square and the road out.

Adam crossed his arms tightly in front of him so as to stop shivering. His teeth chattered. But he was proud he'd done what he'd done. He'd dared. To deal with the policeman twice in one evening was a shock, but he'd made a fool of the man. The policeman knew now what an imbecile he was. And Adam had managed it despite Jenny who was urging him to be a coward. Kissing her was a master stroke. Adam unclasped his arms and slapped his knees. His head was woozy. He looked over at Jenny's prim remoteness behind the wheel, her scraped-back hair a straight line against the paler dark of the car window. How old was she? Her chin wasn't flabby. Would she look sexy with her hair down? To see her straight grey hair fall out of its tie down her back would daunt him. He'd caught sight through a doorway of an old village woman with wanton tresses of white hair. He had turned his head and run. Naked old people frightened him. But Jenny wasn't that old.

Their dinner together had turned out to be quite pleasant. They'd avoided the subject of the policeman, for he feared she might be angry with him and was in no mood to be scolded. Nor was she in a mood for scolding. She had ordered a second bottle of wine, so intent had she been on divulging to him how she wished to become a painter. At home she would draw when Bill wasn't there. Anything, the salt and pepper on the kitchen table, the wing chair in the drawing room. Adam hadn't minded listening.

'I hope he forgets about us. Do you think he will?' The car lurched into a pothole. Jenny turned hard on the wheel to miss where the road was washed away, scratching the bumper on a boulder. It was a dark night, the sliver of moon giving no light: only the car's headlamps showed in a myopic way what was in front of them.

'It taught him a lesson.'

'I hope you're right. Why don't policemen like lovers?' Is that what Bill and Mary Elizabeth had been doing when the policeman wanted to shoot them? Jenny clenched her teeth and gripped the wheel. Why couldn't she and Bill at least be nice to each other?

'Lovers threaten their authority. Lovers aren't afraid.' Adam thought he'd dared to speak up because he was in love. Laura gave him courage. He must tell her.

It had been that morning when he had finished laying out Christopher's new trenches and missed Laura because she'd been switched with Annabel, that he had dreamed of their next meeting. It would be different from last week when Christopher interfered and the German plopped the frog in front of the schoolmaster. They'd bump into each other by accident. Laura would tilt her head; her cat-like eyes under those wide-apart eyebrows would realise suddenly that he wasn't what she thought but an exciting grown-up man. For an intense moment their eyes would meet and they'd know. They'd arrange a rendezvous up the valley, perhaps as far as the old village which was a favourite spot because of the fig tree. After dinner. He'd take a torch and a blanket. She'd be impressed he'd remembered a blanket. Or would she laugh at him? No blanket then. They'd embrace by the cistern where the limbs of the fig were a white ecstasy of entangled limbs in the moonlight. They'd lie down on the bare ground.

Would she marry him? He'd want her to. It didn't matter that she was older. She'd be grateful that he could keep her young by making demands on her that an older man would not have energy for. He'd fuck her four times a night, and demand that every bit of her love him back. No funking it, no grumpiness and coy atrophy like his mother who was as coy and frigid as a rose, her well-lotioned hands and smooth pink face in love with her efficient kitchen. Laura would never be like that. She'd already had many lovers. Adam could tell. She wasn't the virginal type like girls he'd known at home, or like Jenny. Jenny couldn't like it. But Laura loved it, her black tangles, the deep laugh and those long legs . . . Adam felt himself burn with an erection, his penis hurting against his tight trousers. Adjusting himself, he crossed his legs and slid his arm along the back of the seat.

'If you're serious you should talk to Bill.'

Adam giggled. Jenny apologised if she'd said something wrong. 'Serious about what?' he asked.

'I thought you'd said you were thinking of giving up architecture and becoming an archaeologist.'

'Right.'

'Well, Bill could be an enormous help. He knows people. And he'd be interested.'

The car whizzed past Vasilis's café. There were still people there. 'Stop!' Adam shouted. Jenny jammed on the brakes. 'I just saw Bill and Laura.'

'Go then.'

Adam leaned back through the door. 'Are you coming?'

'In a minute.'

CHAPTER 17

Again Laura found herself at Vasilis's on Saturday night eating the same greasy omelette and washing it down with Johnnie Walker bottles of pink wine. She looked about nervously when she and Bill stepped up into the bright room, her eyes flicking from group to group for a glimpse of red hair. But all the heads bent over cards or dice were black-haired or white-haired, the din of men's voices so deafening that with relief Laura led Bill back out to the same table where she and Christopher had eaten the week before.

She hoped Bill didn't notice the leer Vasilis gave her when he brought the wine and bread. She checked; Bill looked as usual, detached and self-absorbed. To have dinner with him would be restful if deflating.

Vasilis appeared with the omelettes and a salad, plonked the plates down on the table and pulled up a chair. He harangued Bill on Britain's slavish attitude toward the Americans and said that Kissinger, not Wilson, was really prime minister. Bill laughed. He told Vasilis that England was not called 'perfidious Albion' for nothing. He had to explain what 'perfidious Albion' meant. Amazed by Vasilis's angry curiosity and the vindictiveness in his keeping to his point as he rested his small grimy nail-bitten hands on the table in front of Bill, Laura began another letter to John. She tried out phrases that would dramatise the outlandishness of Vasilis and the awful omelette he served before he embarked on the politics.

A finger ran up the back of her neck. Laura froze. She stared desperately ahead at the other two, frightened they might see. Michaelis moved away, his long torso and short bow legs blotting out the light from the café door as he faced Bill and Vasilis. Michaelis listened. Vasilis railed against all governments but Bill remained cheerful. Michaelis's thick arms covered in red hair spread across the table and alarmed Laura. She skipped up the two steps into the café to fetch another glass from the counter at the far end. The men watched her walk across the litter of peel and shells and squashed chips, their eyes running down her thighs to her narrow ankles and the red sandals. She gave them all a wave on her way back out, and set the glass between Michaelis's arms. He didn't look up; he knocked the glass against the table and swallowed the wine. The argument continued. Michaelis said that the British government had no choice but to do what the Americans told them to do. Laura slumped down into her chair. She saw that Michaelis would not give her away.

The novelty wore off. Vasilis was like a precocious child who went on and on, careless of everyone else. Bill's laugh sounded less spontaneous and Michaelis's head lolled. Laura had trouble not losing herself in admiration of his beautiful mouth. She kept her eyes down to begin with. But then to tempt him if she could, she leaned her head back and stared up into the clear night. If she looked long enough she'd find a star that fixed her attention.

Had he fucked her to please Vasilis? That ugly thought came back.

A white shirt with a scrawny neck materialised out of the dark. The man tapped Bill on the shoulder and asked if he knew what was in his sack. The man had a long sharp nose that jutted out of his bald head like a broken twig. He was over eighty; his old skin hung in folds. He nudged Bill when Bill couldn't guess.

'Gold,' shouted Michaelis. 'You found the king's tomb!'

Bill sat up straighter. 'What king?' He asked Vasilis to bring more wine, but Vasilis wouldn't go, he shouted to someone else.

'The king,' exclaimed Vasilis. 'You've found his palace, why not his tomb? He died, didn't he?'

Bill eyed the old man's sack, which delighted the old man. A boy brought more wine and Michaelis filled Bill's and Laura's glasses. Wide awake now, Michaelis wagged a thick finger at the sack and asked Bill if he could tell if it held the king's gold.

Bill pressed his hands to his chest. 'Who knows if there was a king?'

'But would you know,' Michaelis persisted, 'if the gold was real?'

Bill shook his head. 'What do you mean by real?'

'Old.'

Vasilis broke in, 'If they built palaces they had kings.'

'We call them palaces,' Bill corrected. 'They could have been monasteries.'

The shock of Vasilis's laugh, which twisted his dark face and moustache into a leer as baldly obscene as if he'd unzipped himself and bobbled his penis at them, chilled Laura. He launched into a story about monasteries; he gave the other men knowing looks as he told it, Laura feeling more and more uncomfortable. Bill vaguely smiled. Michaelis and the old man hooted with laughter, falling about on their chairs. Laura tried to catch Bill's attention. But Bill was not thinking about her; his eyes were on the sack.

The laughter died down. The old man picked up his sack. A jumble of bones tumbled out when he emptied it at Bill's feet.

Adam leaned against the wall of the café in the dark where they couldn't see him. The café was empty although the light still glared, the grinding hum of the generator an abandoned sound like the idling of an empty car. Laura and Bill, Vasilis, the red-headed chap and the village quack sat in the semi-darkness outside round a table of donkey's bones. They were arguing in Greek; Laura would ask Bill to translate. She sat hunched over her lap like a tired traveller. She needed him. How could they ignore her? Adam would surprise them.

'Were we worth spying on?' Laura asked when Adam slumped into a chair beside her. He winked, which made Laura shift back to the bones.

Adam squeezed Laura's arm and whispered that he had something to tell her. She laughed. Adam asked her what was so funny and gazed into her eyes.

Bill said goodnight and left, limping slightly as he disappeared into the dark. The village quack shuffled off into the night with his sack of bones. Laura advised Adam to go to bed. Adam shook his head. Laura pulled him by the arm. 'Come on.' She tried to force him to go.

Didn't she know he hated to be forced? Hadn't she guessed he was the sort who refused? Laura nodded to the redhead to take his other arm; they lifted him and marched him between them down the street. It was flattering that she cared.

Jack and Susan zigzagged up the street ahead. Laura called, beckoning to Jack to take her place. Jack gripped Adam so hard under the arm that it hurt. Adam winced, clenching his teeth. Only Jack held him now; the redhead too had gone.

'Hell, no!' Adam pushed Jack away. Jack cursed, and Adam was left by himself. He swayed on his tired legs. The street was empty. Where was Laura? A glimmer of white shirt floated in the distance. He heard her laugh. She'd betrayed him. The bitch.

Vasilis told Christopher about his cousin in Ierapetra who rented cars to tourists. He had drawn Christopher a map. Christopher pressed an electric switch and two warbly notes chimed inside. They stared at double wooden doors carved with flowers and a lion with an elaborate wrought-iron grille over the frosted glass. Finally the door opened and they were informed that Vasilis's cousin was asleep. The wife had a pretty face, very young with long hair and a smooth complexion. When Christopher did not insist that she wake her husband, she grinned and invited them in. She sat them down at the table just inside the impressive door and made coffee. Christopher didn't mind waiting, she was so lively and thin and graceful in a loose, flowery dress. She gave him a mischievous smile when he explained that he and Annabel were archaeologists. Christopher avoided Annabel's eye. Snores from the husband came through a curtain on the right; the wife pointed, pressed her head to her hands with a contorted expression on her pretty face and skipped off to wake him. 'She could have done that in the first place!' Annabel whispered. Christopher shook his head.

The husband, when he emerged rubbing his eyes, was twice his wife's age with an ugly wart on the side of his nose. The keys were in his pocket and the car, he waved, outside in the street. He led the way. Christopher handed him a 500-drachma note as down payment without flinching, prepared to be at the man's mercy since they needed a car more than the man needed their business. The owner revved the motor to show that the car worked. Puffs of black exhaust blew into Annabel's face, which alarmed her as she coughed and rubbed her eyes, afraid she'd lost a contact lens. She made sure that the lenses were still in place, blew her nose and tried to forget the delay and the criminal price Christopher had agreed to without an argument.

The light as they crossed the isthmus was enchanting, the wall of blue mountains to the east implacable and exciting behind the lush green of the orange groves and the small white churches at Episcopi. Annabel began to have her good time.

They parked below the hill of Gournia and walked up the paved Minoan street to the top, peering over the low walls of the settlement which had been excavated years ago by two American ladies. As they scanned the ruins of neat blocks of houses girding the hill, Christopher told Annabel about the great Harriet Boyd who on the first morning of work when all the workmen were called either Lefteris, Manolis, Vasilis or Andonis sent them all home with orders to return the next day with surnames, which they did. Such assurance was a tale well suited to this peaceful afternoon. Annabel laughed her raucous laugh and strode on.

The sea was a truer, darker blue on the north coast. Annabel exclaimed; Christopher nodded. It was some way to the turn for Kritsa. Long shadows stretched across the road. They drove through acres of olive groves, the rust-coloured earth at the feet of the dark trunks ploughed like a field in England. The church of Kera was locked. Christopher agreed that the wall paintings inside were the main thing but he took Annabel round the back and pointed out its quaint, whitewashed buttresses. The ring of cypress trees protecting it from the world pleased Annabel more. She launched into a description of the long avenues of cypresses in Italy. Christopher left her. She lit a cigarette and rested against one of the trunks. Christopher was too restless to stand still. He walked back to the car.

It was dark when they reached Kritsa. Christopher was eager to find somewhere to sleep, but Annabel had spotted a lace shop that was still open. She got out and he went to find rooms.

Sitting out in the street for drinks was pleasant enough, but the place they found to eat in was a cold high-ceilinged room lit by two bare light bulbs dangling from frayed cords. Out behind, Christopher found the owner's wife standing over a pot of okra and tomatoes. He ordered two servings, salad and wine; and he and Annabel settled themselves at a table against the wall, able to sit wherever they liked since they were the only people eating. One other man sat at a table in the middle under a light bulb to read the paper. Annabel lit a cigarette.

'Tired?' she asked, blowing smoke out through her nose.

Christopher stretched out his legs and yawned. He apologised. Yes, he was pretty tired but he was glad she'd suggested they come to Kritsa. It was good to get away.

Narrowing her eyes like a cat, her head well back to keep the smoke out of her eyes, she asked Christopher if he ever talked about himself. 'I think you're afraid to. Am I right?' she demanded. 'Probably you think it would bore people and you're afraid of being boring. Aren't we all!'

The wine came, a bottle of Gortys red, along with the okra stew which was lukewarm. Christopher shovelled in his stew. Annabel poured the wine, surprised by Christopher's boorish appetite. She watched him while she finished her cigarette. His impassive expression as he ate was like an animal's, not a dog's either but a wilder animal's; it was the expression of a frightened thing eating on the run. When he'd finished he mopped up his plate with bread; his jaw chomped at the crust unintelligently, his face a blank that inexorably consumed. He took a sip of wine when his plate was clean and looked up. 'That's better,' he commented, resting at last. He noticed then that Annabel's stew was congealing on her plate, and urged her to eat it before it was stone cold.

'You were obviously hungry,' she remarked.

'Yes, and I hope he hurries up and brings the salad. Would you like cheese afterwards? He might have some *graviera*. When he comes I'll ask.'

Annabel tried a little of the okra stew and put her fork down. 'In London you never mentioned Bill. He was never at your parties. Is he a very old friend?'

Christopher refilled their glasses and leaned back in his chair. He pushed his chair back from the table so that he could cross his legs. 'I don't know why he never comes to my parties. He's always busy, and he lives in the country.'

It was the wooden floor that made the waiter's footsteps sound hollow as he approached with the salad and more bread. Christopher thanked him and asked if he had *graviera*, holding up the bottle of wine for another. The man was old but he moved with stern determination in his black polo-neck with a crumpled blood-red apron tied around him. Without a gesture or a word he went back to the kitchen; Christopher was left to guess whether he had *graviera*.

'Are you satisfied with your life in London? I know how you collect clocks. But is that enough? You always strike me as a remarkably talented person who hasn't enough to do.'

Christopher smiled and twiddled his empty wine glass. 'I find there's plenty to do. Too much. I write my book when I can and often spend the morning in the British Museum. Occasionally I meet a friend for lunch at that Pizza Express in Coptic Street. We've met there, haven't we?'

'Very art nouveau,' Annabel quipped.

'Good pizzas. And in the evenings I go to my club, when I'm not writing . . .'

'Are you happy?' she interrupted, losing patience.

Christopher was surprised but not alarmed by such intense interest in him. Even pleased. And when the owner came with a plate of *graviera* he was overjoyed; he'd given up hope. The owner had forgotten the wine, but Christopher was much happier now. He offered the cheese first to Annabel who took one of the long, soldier-like pieces piled on the small plate.

Certainly not alarmed. An Annabel could not alarm him. She was too much of a freak, with all those rings on her fingers and her puffy, ugly face painted like a clown's, to be taken seriously. She was pitiful. He could say what he liked to that ridiculous black scarf tied round her head and the décolleté jacket with its puffed-up sleeves looking like an ancient granny in party dress. She was really rather a relief, fussy and harmless. And quaint.

'That's not a question I ever ask myself,' he answered.

'You should.'

Christopher tucked into the cheese; he pulled apart another chunk of bread. Finally more wine came and Annabel was pouring it out. 'You should,' she persevered, 'because then you'd realise how unhappy you are and do something about it.'

Christopher stopped eating to take this in. With his mouth full of cheese, his eyes flicked up to the high dark rafters lost in shadow. The peeling blue walls at eye level were awful, and haphazardly hung with the Aroma advertisement of the girl on the rocks, and on the wall opposite was a tray from London with a print of Tower Bridge. Cheerless places these restaurants were, like most provincial bars the world over. The man in the middle of the room still read his newspaper while his poor wife probably wondered when he'd come home. Maybe she was resigned. Christopher sighed. Outside in the street a man shouted; inside Christopher could hear the owner's wife shouting threats at her husband. It was the end of the day when everyone was tired.

Oddly, Annabel now was leaning across the table. She'd heard his sigh. She smiled at him, and his hand which had been around his wine glass was now in hers. Had he let her? It didn't matter. Her face was close. He saw her lipstick was gone, her pale mouth giving her theatrical appearance a frail look. The blue eye shadow and black eyelashes were still thick but her ordinary face began to show through. It moved him. He leaned over and kissed her. And sat back down again, pouring out the wine. Annabel looked pleased; they clinked glasses, giggling at the joke of it all.

When the owner announced that he wanted to go to bed, Annabel suggested they go somewhere and talk since it was only ten o'clock and they didn't have to get up at five the next morning. But outside the street was empty and only the street lamps were still on. The cafés were shut. They walked up the street; Christopher took Annabel's arm to steady her, the uneven cobblestones and low kerbs difficult to see. The balconies overhead shut out what light might have come from the bright sliver of moon. 'Let's go back to the hotel,' she said.

Christopher agreed. Everyone had gone to bed.

'I have a bottle of brandy in my room. We can talk there,' said Annabel, unhooking herself from Christopher's arm to walk ahead.

The owner of the hotel scolded them for keeping him up. Their keys, which he had thrown down with contempt to make his point, clattered on to the desk. Christopher apologised. As his nanny used to do when he apologised and she forgave him, the owner grunted ungraciously and shouted after them a gruff good night.

Annabel sat on the bed and Christopher sprawled on the only chair, the only glass in the room shared between them. There was a bare tile floor and an orange *flokata* on the bed, and the walls were papered in blond imitation wood. Annabel had taken off her rings and her scarf, and unbuttoned the top button of her jacket. Christopher tried to talk but his eye would wander off to corners of wallpaper that had come unstuck; one corner curled up just over Annabel's head. 'I must go to bed,' he muttered, gulping down more brandy.

'Don't. I need you here with me.'

Christopher frowned but before he could ask why Annabel was leaning over him, she kissed him. Then she ran her hands down his shirt, and squatted between his legs. She felt down his thighs, behind his knees, the calves of his legs, his ankles. Her hands gave infinite relief. Idly he fondled her hair, succumbing blissfully to Annabel's massage. He shut his eyes. His low white bookshelf and the blue and white striped curtains at home that smelled of chocolate drifted through his mind; his collection of shells and the model of HMS *Victory* shimmered precariously on the edge of sleep. Christopher groaned. The pleasure! Annabel took off his shoes, then his socks. 'You shouldn't be doing this,' he mumbled.

Now her hands moved back up his leg and he opened his eyes,

suddenly aware of what was happening. But how extraordinary. When he tried to stand up she held him down. She was unbuttoning his trousers and with outlandish temerity cradled his penis. Should he laugh? He could only see the top of her head. He gripped her shoulders and pushed her back. She looked up and grinned.

'All right,' he shouted, 'if that's what you want!' He yanked at her jacket which she shed before he could rip it. There hung her huge breasts. Like a blind man he pushed his face between them and she pulled off his shirt.

It was she who stopped him from going further until they were on top of the orange *flokata*. He was violently impatient. He pushed her down. Had she guessed he didn't know? Had she guessed? He was in. He'd slipped in somehow. She groaned and he groaned and it was over, the long hairs of the orange *flokata* tickling their legs.

Annabel was delighted to wake up in a strange room. Nothing about it was familiar now in the daylight. She threw off her covers, impatient to see where she was. The WC she found off a small porch where there was also a cold water tap. To be brushing her teeth in the sunlight with a view of terraces and hills and a dirt track that curled down to the sea was lovely; a basil plant on the ledge at her elbow scented her fingers. From her bedroom window as she dressed she could see down into the main street where she and Christopher had walked in the dark. The agitated ring of a tinny church bell made a Sunday sound although the empty street gave no indication that anyone was obeying the summons. But it rang and rang.

Annabel wanted to explore. The orange *flokata* which she'd found too hot in the night lay in a heap on the floor. She left it there. Later when Christopher appeared she'd know if he wanted to carry on. Men his age were unpredictable. She hurried out into the street, and met a few women with small scarves tied round their heads. She fell in with them.

But Christopher lay under his covers as still as a corpse with his eyes open. The sun outside his window hit the ochre building opposite. Thank goodness he'd come back to his room. At least he was alone. That drab mustardy sheen outside was right. Ochre had always been his least favourite colour.

Christopher turned on to his side and pummelled his pillow. On his back again, he lay corpse-like as before. His head ached.

The small back porch was in familiar disrepair, the plaster coming away from the wall. The creak of the rusty tap tired him. He leaned on the ledge beside a basil plant and watched the ribbons of mist between the hills rest in the sunlight.

Annabel waved from her table across the street. She was sipping a coffee in her denim skirt and a winsome blouse buttoned up to her chin. Christopher stopped to look up and down for cars before he crossed the street.

'The car's got a flat tyre. But have some coffee first.' Her rings waved him to sit down. Her face was freshly painted; two sharp peaks of lipstick glistened. Christopher sat down. 'I've spoken to several people. A young policeman fetched his bicycle pump and tried. But there we are. Have your coffee. We've got all day.'

The café owner brought Christopher a double *metrio* and a glass of water; Annabel returned to her book.

'What are you reading?' he asked.

Abruptly she held up the book. 'Pashley. A reprint. Gorgeous etchings.' She pushed over the picture of a Turkish mosque in Rethymno with the peak of Mount Ida in the background. 'Crete must have been marvellous then!' Annabel slapped the book shut. 'It's a terrible bore about the car. That damn cousin of Vasilis's is a crook. I hope you're angry, because I am.'

Christopher apologised.

'That's not the point. You've been here thousands of times I know, but I haven't and I probably won't ever get here again. I'll never see Lato now.'

Christopher tried to keep his eyes off Annabel's white frills, recalling how enormous her breasts were. Under the table there were also the fat sunburned thighs and black pubic hair which he'd had under him on the *flokata*, covered now in blue denim like the statue of a King's Road priestess. Christopher studied his coffee. Should they take a picnic to Lato? Cheese, bread, tomatoes, maybe grapes, wine, would be enough.

When Christopher pulled a spare tyre out of the boot of the car and showed no surprise that it was there Annabel was miffed. He might have told her. The car was such a wreck she hadn't even looked. She glared at Christopher's green back in the same boring shirt, bent over the jack which he'd also found in the boot. She'd meant to forgive him for his clumsy effort in her bedroom and even be pleasant so that they made the best of this Sunday which they must spend together. She'd heard back in London that he was shy with women. But a virgin at fifty was no joke! Who else would have pretended his hulk squashing her and gasping like a fish was normal and satisfying? Now to be deceived by this silly fart, her morning already ruined with worry, was intolerable. How could she have ever fancied such a jerk! Annabel turned away and walked a few paces up the street, deciding to buy the embroidered tablecloth she'd seen the evening before. Bother the expense. 'I'm just off,' she called, careless of whether he'd heard or not. She strode up to the plane tree that blocked the street at the top of the hill, to the little shop tucked behind its whitewashed trunk.

CHAPTER 18

Her second Monday. It was a dull day, with pale clouds obscuring the peaks of the mountains, as Laura walked up the valley to the cemetery hill in her blue anorak. It wasn't hot, but not cool either; it was colourless. The olives looked grey and dusty, the deep green of the orange and lemon trees drab in the early, flat light. It was strange always to be up so early. Never before had Laura seen dawn after dawn day after day when it wasn't the end of being up all night. It surprised her that she could manage it; never an early riser at home, she wondered if it were changing her into a different sort of person. A less impetuous, less erratic, less emotional person who could be steady and efficient, and accomplish masses before the rest of the world woke up. Could she clean the house before breakfast? Be bathed and dressed by eight o'clock? It was amazing how much of one's life one lost in sleep. But then what would she do with the rest of her day after eight o'clock, with endless empty hours to be tackled and made something of? Laura sighed. Ahead Susan and Edward walked either side of the road, a good three yards apart, and didn't speak. Right up in front were Andonis, Georgos and the little Manolis. No Michaelis. Her gut lurched as she looked. When she couldn't see him she was relieved. Her quiet weekend in the village abandoned by everyone had turned into something quite unexpected and she needed a rest from the man. She'd given in to him utterly, which perhaps was wrong. Perhaps yesterday she should have said no. But on the other hand he was very sweet, harmless and seemed keen on her which was nice. Love in these lemon groves was a lark!

A hard, even step caught up with her. Christopher overtook her without a word, eyes on the ground, looking dishevelled, the collar of his green golf shirt half sticking out of his khaki jacket. Was he ill? Not even a 'good morning' as the grumpy, hunched figure lunged ahead in his heavy brogues. He bumped into Susan and didn't apologise, which Laura could see annoyed Susan when she had to step aside. He didn't notice. He walked on.

'I'm sweeping again, am I?' Laura teased him when they reached the top of the hill and Christopher ordered them to different trenches. He'd obviously decided in advance who would go where, but was still angry about it. He made everyone feel guilty. Was he sorry he hadn't more supervisors? Was it Annabel he missed? Had he and Bill had a row? Was he in love with Annabel? How amazing that would be, but then he was old, with taste enfeebled by years of abstinence. He'd skipped supper last night which wasn't like him; only Annabel appeared and behaved, in fact, more or less as usual. Perhaps Annabel had rejected him and his love was a hopeless love, yet again. Were all his loves hopeless loves? What a wasted life, poor chap. Laura skipped off with a broom and a dustpan to obey orders and prepare Susan's tomb for photographing. Laura was sick to death of sweeping but obviously Christopher was in no mood for complaints.

A peace descended. As the sun rose higher behind the clouds and the daylight brightened there was only the creak of the wheelbarrow shuddering across the rough ground to the dump, and the scrape of shovels. No one talked, even Georgos was too intent on his work to crack his usual jokes about the greed and laziness of the local villagers (a favourite topic with him). Every now and then a bird perched in the olive tree and twittered in the still morning. Once a flight of swallows swarmed the tree and made a commotion before they moved on.

By eleven o'clock, when everyone stopped for *kolazo*, Laura had finished sweeping. Her back ached from bending over and there was even dust in her mouth. She refused her bread and cheese, drank down several cups of water and found flat ground where she could stretch out and shut her eyes. It had better be soon, she was thinking, that she did something less menial. She'd watched Susan keep a notebook, impressed by her speed in describing features, her certainty like that of a professional detective. If only Christopher would give her a chance. After all, people always remarked on how observant she was, were even sometimes put out when she noticed things about them which they'd hoped weren't noticeable. Should she speak up and not wait to be asked?

Christopher did a last sweep himself to make sure the brush strokes were straight: they mustn't cross or change direction. He borrowed Andonis from Susan to bring over the tall stepladder for a high view. Christopher cautiously hung the straps of both cameras around his neck. By now it was hot. He'd taken off his khaki jacket, the collar of his green golf shirt turned up to protect his neck. Laura sat cross-legged above the *dromos*, surprised Christopher wasn't in his usual denim 'dig' shirt. Was he behind with his washing, she wondered, amused if in fact it was that mean little pile reproaching him from the floor of his room which made him so disagreeable this morning. Washing could get one down, and the green golf shirt looked awfully good.

When Andonis had erected the ladder across the end of the *dromos* Laura stood up to help hold. She yawned. She was almost asleep when Christopher nudged her to wake up. What an annoying boss. And how ridiculous he looked now juddering like a jelly on the rickety ladder. The perfect photo for the records was all that counted. Laura yawned more; she couldn't stop. So much recording; there was no life or excitement in recording. Trap and kill the evidence; they were like hunters, these archaeologists. Laura laughed. Of course they weren't that bad, but they could be very boring. Christopher was finally satisfied and climbed down; he prepared to take more photos close up and knelt to peer in at the brushed walls of the empty chamber.

'Run,' he shouted at Andonis. 'Bring me a small pick. Hurry!' Within seconds Andonis was handing one to Christopher. Christopher on his hands and knees had already messed up Laura's beautifully swept dust. His long bulk leaned into the chamber to pick at a patch of darker earth on the bared 'floor'. The earth of this patch was damp, he mumbled to Laura who leaned over his shoulder. It was the slightly darker colour that made it show up; he showed her how easily it came away from the hard floor. 'Tell Susan to come.' Christopher waved Laura away.

Christopher's pinched face exuded misery as he climbed clumsily out of the *dromos* to wait for Susan. He looked more forlorn than Laura had ever seen him. 'What is it?' Laura asked, out of breath from running. 'Whatever's happened?' Christopher ignored her question. He gestured to Susan to follow him down into the tomb.

Laura held the tapes. Susan's small hand made infinitesimal moves of the pencil as she drew the earrings and the pin in their socket of damp earth. Christopher was adjusting his cameras. No one spoke. When Susan finished he crouched with his flash bulb at the entrance to the chamber. Then he reached in and lifted the things, clenching them in his fist as he backed away and straightened up. 'You'd better get an envelope,' he told Susan in a dull voice.

They were beautiful. A pair of gold bull's head earrings and a gold pin. The heads of the bulls were large and elaborately worked, a wide intricately sculpted jaw balanced beautifully with the horns which curved back around the skull to crown the animal. So delicate, so skilful, so sophisticated. Susan held them, then handed them to Laura. The gold shone. They were surprisingly light and as yellow as a daisy. Awed, the others gathered round and stared at the three small, priceless objects in Laura's hand, still as brilliant as when they were made thousands of years ago. There was an eternity in gold, the immortality of the soul a golden idea, Laura decided. To live on and on was a hope that such undimmed brilliance could provoke.

Christopher's mood hung over them as he disengaged himself from their excitement. He rejected their surprise and joy as a distracted father might do, too fussed with his cameras to look. Laura handed the gold back to Susan and asked Christopher what he wanted her to do. 'Sweep, so I can finish.' He pointed to the marks his knees had made. For Christopher to spoil such a moment! Laura wanted to hit him and kick him and throw rocks at him until he changed. It was wrong to be so distracted and glum. The others, even Georgos and Andonis, had noticed and looked sad now. That first moment when they laid eyes on the gold their faces had beamed like children and Andonis had shouted at Christopher that he was the best gold digger in Greece. Christopher just shrugged.

Work went on. Manolis had started to uncover a tumble of *kouskouras* which were the chunks of white marl from a collapsed roof. Another tomb. Lower down the hill Edward and Georgos were finding a few sherds of Late Minoan bowls but the earth was loose, which meant the tomb had been robbed. Susan and Andonis were removing their surface level still, their trench only begun that morning. Again they'd forgotten to bring oil for the wheel of Andonis's barrow which as the hours passed became more irritating as it squeaked across to the dump.

But profiles of jars and plates appeared as Manolis cleared away the *kouskouras*. It was like watching a negative in the developer slowly reveal what it had. Christopher worked alongside, sweeping as Manolis picked.

Suddenly Christopher straightened up and jerked his head from side to side, as if he'd that moment put something down and couldn't see where. 'Get Georgos,' he ordered Laura.

The tension thrilled Laura as she ran down to Edward's trench and told Georgos to come. He ran back with her and Christopher told him to walk to the other site as fast as he could and tell the director to bring the foreman.

Christopher stared after Georgos. His gaze crossed Laura's.

'Please cheer up.'

'What do you mean?' he snapped, frowning at her. Laura pinned back some wisps of hair and waited for her next orders. What a pity there was no Michaelis nearby. She was missing him now. It was Christopher's fault.

By the time Bill appeared Christopher had Laura on the run fetching labels, bags and zembils so that when he'd drawn and photographed everything she could tie on labels and pack the pots carefully in the zembils. Andonis was put to shovelling and barrowing, summoned by Christopher to help out when more and more pots showed up in the packed earth.

Christopher handed Bill the envelope with the gold earrings and pin which he was keeping buttoned up in his top pocket. Badger took from Christopher his small pick and joined Manolis in the trench. He pushed the boy aside.

Bill turned the earrings over in his hand and slid them back into the envelope. He handed back the envelope and moved to the other side of the trench where Laura was tying on labels. They were lifting the things. Laura jammed grass, twigs, whatever there was between the pots as she packed them into the rubber baskets (which she congratulated herself on knowing to call 'zembils'). 'All properly labelled? You've been shown?' Bill asked her. Laura ripped a branch off the olive tree and wedged sprigs of that between the next three plates and a stirrup jar (again she was pleased with herself to know to call it a 'stirrup jar' with its stirrup-like handles). She glared at Bill, angry that he thought she didn't know what she was doing.

Long shadows stretched across the tomb now. Susan and Edward had gone down hours ago but Georgos waited with the donkey. He and Andonis and Manolis were still there. And Bill. Not once had Christopher given Laura a smile although she had not stopped labelling, packing, labelling, packing. Georgos had already taken two loads of her carefully packed objects on the donkey down to Bill's car.

Bill picked up a shovel and smoothed over where Manolis and Badger had finished for the day. Christopher pulled out the last thing, which was a model of a bird. He held it up so that Andonis could see. Laura by this time was so fed up with Christopher that every action of his seemed foolish. To make such a favourite of Andonis was typical of his unaccountable behaviour all day. Only that old workman was to be graced with a smile. While the rest of them slaved away, the bastard! Laura was eager to get to her bed.

A few steps from her house, an hour later, Michaelis jumped out from behind the pile of rubble and hissed at her, 'You think you'll find more gold?' Laura squealed. She nearly fell on her face, he'd startled her so. He laughed and grabbed her arm. 'I'm taking you with me up into the mountains. I've been waiting,' he said. He smiled as if he'd done nothing wrong.

Laura remembered herself now, afraid of Annabel in particular. All she could see was the back of an old woman pulling her goat a few yards away.

'What are you doing here?' she whispered. 'Go away!'

'Waiting for you. We go now.' He squeezed her arm hard, which frightened her. She pulled away. He blocked the way, with feet apart and arms akimbo. 'We go now,' he ordered.

Laura rocked precariously on some stones, aching with fatigue. It was after six. The twilight was an icy blue, the white front of her house shimmering in the dim light. 'I'm very tired,' she sighed. 'It's far too late to go to the mountains.'

Michaelis shook his head like an obstinate child. His brown eyes were glued to her face. She felt like a nervous schoolgirl all of a sudden, confronted by this Cretan lover. She giggled, more pleased than frightened, by this time.

Michaelis marched off down the rubble heap to the street. Laura called to him but not loudly, afraid of Annabel. He didn't look back. He had not heard. Laura sighed, too tired to mind if she'd hurt his feelings. She'd apologise sometime. The silly man hadn't even let her tell him about the gold earrings, and how Georgos was guarding the tomb.

Laura collapsed onto her bed. She didn't have long. She was to meet Christopher to go with him to the police station. He and Bill had had an intense discussion at headquarters just now, when they were finally alone, about what it all meant. Others looked in to see the earrings when they'd had a swim and changed clothes. At last Laura witnessed what she had missed up on the hill – how finding gold put them all in a fever. Even Bill, when there was no workman to see, sat like a proud father before his team's find. The gold bull's heads lay on the table like a family treasure. Christopher leaned his head on his hand and stared. He wasn't eating. He'd poured brandy for the three of them from his flask and refilled his glass several times.

When Christopher asked Laura to come with him to the police station, Laura was delighted. What a compliment! She forgot how angry she'd been with him.

He would also, he told her, telephone the Ephor from the police station and give him the news.

Christopher was tortured by a picture of himself with his head pushed between Annabel's breasts and his erect penis forced up her. He was a filthy forty-eight-year-old. He'd shocked himself. He was afraid now of his lust and his anger. In that awful hotel room, if he hadn't flared up in anger, he wouldn't have pressed himself on top of Annabel on that horrible orange *flokata*. He couldn't love. He'd never learn. He could only attack. The anger gave license to lust and there he was on the orange *flokata* raping that poor woman! Thank goodness she'd asked Bill to switch her back to the other site.

He was sitting at the café by headquarters waiting for Laura. A sadness had set in like the guilty depression of a hangover. It wasn't fair that he should find gold and an unrobbed tomb. His good fortune was an embarrassment. It had foisted on him what it knew he could not enjoy because he knew he didn't deserve it.

That he and Annabel had got through Sunday was her work. He'd bought them a good picnic and she'd been pleased to see Lato. She pretended nothing had happened. But as the day dragged on he couldn't ignore the stark truth and lost his appetite under the oak tree where they settled, after walking about the hill studying the grey ruins and the long views down to the sea. Annabel never mentioned what he'd done so he couldn't apologise. But anyway who could forgive a middle-aged hypocrite?

When he and Annabel got back to the village he went straight to his room and lay down. He skipped supper. Then there was the knock on his door and an anxious Jenny with a note. That was the last straw. The note was brief; Bill presumed Christopher realised that the Ephor 'favoured' him and warned Christopher that the village thought he was finding the king's tomb. Was he prepared for that? Perhaps he deserved that note. He tore it up all the same when Jenny had gone, and cursed Bill for distrusting him and disturbing his sleep.

How lucky for Laura that he didn't rape her. He nearly did. Christopher got restlessly to his feet. Laura was in a pretty red and white striped dress when she finally came. Christopher pulled his hands out of his pockets and smiled for the first time that day.

Christopher congratulated Laura on thinking to put on a dress since the police were not keen on women who wore jeans. Laura thought, how pompous, but brushed it aside as they set out together. Bill had just come round the corner at the other end of the street. They waved and walked on to the police station which was in a new building on the edge of the orange groves. Bill had refused to come with them, still in a huff at the Ephor for avoiding him on Friday. He made Christopher do everything.

The police hadn't yet taken down their flag although it was after sundown. Two men in uniform were leaning against the railing of the verandah when Laura and Christopher started up the steps. They stared hard at Laura and gave Christopher a cursory glance. But Christopher ignored the impudence and suggested they go inside to talk. The men in uniform were slow to follow. The older man walked to the far side of the large table in the middle of the room and sat down. Christopher laid the envelope down in front of him.

The older of the two policemen was completely bald with ears that stuck out like fins. 'We've been waiting for you,' he reproved them, covering the envelope with both hands. He was unshaven, the grey stubble strange on the face of a man with such a neatly clipped moustache, his high collar buttoned right up to the chin so that it forced a fold in his jowls. The other policeman's jacket hung open, revealing his vest and hairy chest. He was much younger and quite good-looking, Laura admitted to herself. The younger one snatched the envelope, chided but not stopped by the older man. The two gold bull's heads and pin slid across the table. Christopher, horrified, quickly put out his hand to stop one falling on to the floor.

'What took you so long?' asked the older policeman, who was obviously in charge. He'd reprimanded his young colleague for his carelessness, sliding the earrings back into their envelope. 'You found these this morning and it is now late evening.'

'We worked late,' explained Christopher. Laura self-consciously looked around the room. On the wall hung a picture of Papadopoulos and under him a photograph of the king. Further along there was a framed list of ten sentences. On a chair was a tray with dirty coffee cups and half-full glasses of water. The room was a void of officialdom, but there was the smell from out at the back of meat grilling, the aroma of charcoal filling the room.

'You are finding the royal tombs?' It was the younger man who persisted in his bumptious ways, unabashed.

'No,' said Christopher.

The young man laughed. 'Of course you are. We all know. He was our king once upon a time. Ours! You understand? You think you'll find more gold. Of course you do. But we are the descendants of this king.'

'Where were you born?' Christopher shot back at him.

'I am a Cretan.'

'Where?'

'Do you know Crete that well?' asked the young man.

'Very well.'

'Krasi. Have you been there?'

'It has a huge plane tree. Very big. It's famous for that.'

The young policeman moved away from the table, impressed by Christopher's knowledge. He disdained more conversation with him.

As the older man was sorting through a bunch of keys, Christopher asked him if he might telephone the Ephor. The policeman said he knew the Ephor had visited the village on Friday, and pushed the telephone across the table. Then, still holding the small envelope, he unlocked the door into a cell. One small window had bars and there were four rifles in the corner opposite a bench. Above was a small cupboard. Another key opened that, where the white envelope was carefully laid, the door shut and locked again with a thud.

The Ephor was still in his office when the telephone rang. He'd stayed late. Loud crackles made it impossible to understand who was speaking. But he did not hang up, not surprised when he finally caught the name.

Much of the day he'd stood at his window and stared out at the ragged palm blown against a dismal sky in the heavy air. His official visitor of last Thursday was a tight knot that still curdled in his stomach and sapped his energy, corroding his gut like a bad meal. The infamous Demetriadis had barged into his office, his bald head and heavy jowls smelling of cologne. The first thing he did was point out the crack in the glass on the photograph of Papadopoulos. It wasn't hanging on the wall above the photographs of the king and queen as it ought, but lay on top of periodicals, staring at the ceiling. 'It needs a new glass,' he'd remarked, tracing the crack with his pudgy finger. Sideris Koraes had known Demetriadis in Macedonia when he'd connived with the Germans. He was a shameless intriguer who needed to frighten and betray to feel alive. Recently the Colonels had snatched him out of retirement to spy on his colleagues, so Sideris knew why he'd come to see him at the museum.

Demetriadis had leaned out of his chair in his navy blue blazer and starched white shirt to warn Sideris that Crete was full of Communists. They were Communists turned robbers who threatened the security of Greece. In particular, he rushed on, there was a café owner on the south coast, a Vasilis Andonakis who was a well known Communist and tomb robber. His impatience implied that Sideris had not kept an eye on the café owner because he was in sympathy, but the government in Athens was not stupid. He shook the cracked photograph at his colleague. He must please explain himself. Any tombs been robbed recently? What did he do? Confess! How did he vote in the sixties? Demetriadis let the photograph drop onto the periodicals again and sat back. He eyed Sideris with such triumph in his piggy eyes that Sideris lost his temper.

To suggest that a fellow Greek and an old colleague would connive with robbers was a shame! Demetriadis dishonoured the Greek nation and the Greek people and the virtues and ideals they stood for, shouted Koraes. He was no Greek! Sideris Koraes loved his country. He loved its past. As an archaeologist he dedicated his life to preserving this past for posterity. How dared Demetriadis call him a tomb robber?

'Then catch him,' Demetriadis had challenged him. He opened out both hands and shrugged. 'You don't want to go to prison.' Afterwards he spent an hour in the Epimelete's office, so his secretary Mina told him, to extract information. Sideris suspected it was his Epimelete who'd informed on him in the first place, a typical mainlander who hated the Cretans and called them bandits, drunkards and Communists. Most likely he also resented his Ephor.

Sideris told no one that Demetriadis had threatened him with prison. Did Demetriadis know of a plan to frame him? Sideris had many enemies in the Ephorate who disliked him for being a Chiot. They thought him arrogant and too much of a friend to foreigners. They envied his international reputation as an archaeologist and his seniority in the Ephorate. If the plan was to frame him, he was done for. Anger and fatalism kept him awake at night, but he did not tell his wife why he couldn't sleep.

To see Bendick's excavations on Friday had been a relief. He liked Christopher, except that he never said what he thought. When Christopher said he knew his workmen, Sideris interpreted that to mean he didn't know them but would be loyal and expected good work. This provoked Sideris to remind Christopher that he was responsible.

Sideris regretted what he did afterwards on Friday. He found Vasilis Andonakis's wife behind the bar at the café and asked for her husband. The boy brought him. Sideris suggested Vasilis walk him to his *thermokypia* (he'd found out from the wife that they had one) and show him what was growing under the plastic. Vasilis asked Sideris what he wanted. 'Don't you trust them?' he'd asked, meaning the English. When Sideris assured Vasilis that they were trustworthy, Vasilis asked, 'What are you afraid of?' By now they were out of the village on the coast road. Vasilis pointed up to his hothouse and Sideris agreed that he'd rather not climb but listened gravely to Vasilis's account of the hard work it was to prepare the ground with beach shingle. 'But Cretans aren't afraid of hard work,' Sideris argued, edging closer to the subject on his mind. This man was proud and had a temper. If he was a Communist and enemy of the government as well, as the authorities suspected, it was like handling explosives – even if the man wasn't brave (which Sideris suspected) he was stubborn, and his cunning made him unpredictable. What was Vasilis's weakness? Besides being evasive, was he greedy? If so, he could do something foolish, being so set on getting more for himself that he ignored the reactions of others.

Vasilis gave Sideris an impudent look. 'What are you worried about? You were here in the spring. You never found out, did you, who robbed the tomb.'

'That's not true.'

'Is he in prison?'

Such effrontery was a penance but Sideris was too frightened by the mention of 'prison' to show anger. Hate of this wretched peasant, however, burned inside him. He'd started out with a cautious respect for the man's independence. But Vasilis went too far. 'Yes,' he lied.

'That's a lie,' said Vasilis.

'How do you know?'

'See? There! You've admitted it,' Vasilis shouted at the tall, distinguished figure of the Ephor who walked with hands clasped behind his back.

Sideris turned on the man. He warned him. He was known – the Communist who had no respect for the law. One more robbery and it was prison.

'But first,' retorted the incorrigible café owner, 'I must be proved guilty.'

'You are guilty!' Sideris shouted. He'd lost his patience, a terrible mistake, he saw now. He'd left Vasilis in the middle of the road with his hands in his pockets. He feared Vasilis might take his foolish threats as a challenge. What if they provoked him to organise a second robbery? Sideris would go to prison. And Vasilis would exculpate himself. That would be part of the challenge, to frame someone else.

Sideris went home to bed. On the Saturday he went to his office to be alone. He pulled out of a bottom shelf his copy of *Burke's Peerage* to look up Bendick. He was right that Christopher Bendick's father was the younger son of an earl, Fitzclarence. He ran his finger down the column of Bendicks, soothed by the security of lineage and the light it cast on Christopher's inscrutable courtesy. Aristocrats behaved differently. It was only a hunch that Christopher would cope better if there were more robbing. Unlike Bill Courage, he was not ambitious and he was unafraid of the Ephor.

Suddenly the line cleared and the Ephor could hear. Sideris asked if they'd kept it a secret. The workmen on the site were Knossians, Christopher answered. He and Bill had ordered them not to talk. The police knew, of course, but he did not think anyone else knew.

Sideris was doodling down the margin of a letter from the ministry. His doodle had reached the bottom of the page. Elfin faces with pointed ears and negroid lips tumbled between spirals beside the new terms of holidays for museum guards. Along the bottom he did figure eights as he tried to decide how much of what Christopher said he could believe.

'Of course it's general knowledge. Do you think I'm stupid?'

There was silence. Sideris had called Christopher's bluff. Now he had to decide whether he would inform Christopher about the suspect café owner Vasilis and describe his talk with the man on Friday.

'Who is there now guarding the tombs?' Sideris stalled for time. He couldn't decide.

'Our Knossian Georgos. And later it will be Andonis Markakis, another Knossian.'

'They were your workmen today?'

'Yes.'

'And you don't think anyone else knows? I am to believe that? All right. I believe you. Good luck. Tomorrow I shall be sending you my Epimelete Costas Pavlidis. He will collect the earrings. Goodbye, Christopher.' Sideris hung up abruptly before Christopher could say anything more. Since he had decided that it was better Christopher should not know about Vasilis, he wanted no more conversation with him. There was no need to reveal how badly he'd handled Vasilis.

Sideris neatened a few of the piles of papers on his long table and picked out from a low shelf a favourite P. G. Wodehouse, *The Inimitable Jeeves*. He leafed through it idly, stopping at Chapter Six: 'I don't know if you've noticed it, but it's rummy how nothing in this world ever seems to be absolutely perfect.' 'Rummy' was such a good word – his mess-up with the café owner was rummy, a significantly nebulous word which made him feel better at once.

For twenty-five years Sideris had been Ephor in Crete. His title was an ancient Greek word for inspector: anywhere else he would be called Inspector of Antiquities. But in Greece he was called Ephor and he preferred it, glad that the title was not an import from another language. Although a Chiot brought up in Thessaloniki and educated in London, Crete was his adopted home. He would never leave now. He loved Crete, partly because he loved its beautiful mountains but even more for the passionate independence of its people, and its past. The peace-loving Minoans had been the academy of the world, as centuries later Athens was for Rome. He was still studying the finds from his many excavations; he had only completed the publication of his first excavation, a farmhouse in East Crete, which saddened him. It was especially difficult now, hectored as he was by the ignorant and venal agents of the Colonels in Athens. He couldn't afford to retire, and he felt it was his duty to protect the antiquities of Crete from tourist operators and the other 'friends' of the Colonels.

A philosophical detachment had calmed him at last, and he was hungry. His wife should be home from the hospital soon, where she was a consultant. He turned out the lights, with Wodehouse tucked under his arm, and locked his office, walking slowly down the hallway that led to their private apartment. He admired the English enormously. They were loyal, and they were stupid with superb style. They didn't fret about what was true. They ignored such questions and believed what they wanted to believe, quite often succeeding in making come true what wasn't really true at all. He chuckled over Christopher's belief that no one in the village would know they'd found gold because they hadn't been told. Was Bendick that naive? Who cared. Rummy it was. Delightfully rummy.

CHAPTER 19

The moment Christopher set foot through the door Bill attacked him. He raved at him. Jack was leaning against the wall behind.

And Christopher couldn't make out a word of what Bill said. He shouted back at Bill to calm down.

'All right, I'll tell you.' Bill's voice dropped to a murmur. 'A little while ago I was passing the Merry Widow's café . . .'

'Which one's that?' Laura interrupted.

Christopher pulled out a chair, sat down, and smiled at Laura. He poured out wine for both of them.

'Is it all right if I continue now?'

Christopher nodded.

'I heard the workmen. I walked in. There was Andonis with everyone sitting round describing every single thing you found today. He even exaggerated. He told them for instance that the gold bull's heads were the size of two big tomatoes. And everyone there believed him. Their mouths hung open with their eyes shining. He was exultant, that wretched chap, and even when I came in he didn't stop. He was shameless, though he was doing exactly what I had told him not to do. Every person in this village knows now. It's spread to Ierapetra. And beyond, to the mountains, Symi, Kato Symi. Every person on this island will know by tomorrow that we've found gold!' Bill spread out his arms. He looked like a crucified Christ. Laura laughed. Smiles lurked on the other faces. Only Christopher remained solemn.

'What should we do now?' Christopher asked.

'It's done.'

'Perhaps we should talk to the mayor.'

Bill gulped wine. Margaret brought in a huge pot of soup, Jenny rose to her feet to serve it. Only the knock of the ladle on the crockery sounded in the room.

'What to tell the mayor?' Bill asked.

'The reputation of the village must be guarded. Everyone must keep an eye out to safeguard *their* treasure. Couldn't we try that?'

'Even the Kallithea site's not safe. They'll think there's gold even there.' Bill rubbed his cheek.

'I don't think so.'

'No one likes the mayor.' It was Adam. He tilted his chair right back so that his head touched the wall. 'He's the Colonels' man. Only the schoolmaster and the German speak to him. And the police, of course.'

Everyone round the table eyed Adam with unanimous contempt.

'What about Vasilis? He's proud,' Christopher pressed. He stuck to the idea that the village would be persuaded to help look out for robbers.

When they'd finished their soup and Margaret was coming through with a chocolate cake covered in gold beads, the sudden creak and scrape of the door startled everyone. Andonis with his bare hairy arms and wrinkled face stood weeping in the doorway. Laura pulled in her chair. Bill stared from the other end of the table. Margaret started to sing 'Congratulations'; she fell dumb with the cake poised in her hands. The shrill of Andonis's childish cry was the only sound. He fell on Christopher and babbled that his son the bus driver in Heraklion had been hit on the head by two drunk passengers and lay unconscious in hospital.

'All he did was ask them to move along,' Andonis wailed. He turned his face to the wall; Christopher laid a hand on his shoulder.

Bill pushed Christopher away and demanded more details from Andonis. Andonis asked if he could go home, but Bill said no since he was needed to guard the tomb. He was to relieve Georgos and watch through the night, which was vital now that he'd told everyone in the village about the gold. Next day he could go, when they'd finished the tomb.

'By then he'll be dead!' wailed Andonis.

Bill was too angry with the man to relent. He ordered Andonis to collect his things since in a minute he would drive him to the hill. Bill walked the old man out of the room. When he came back and shut the door no one spoke. Margaret gave Susan the cake, and she cut everyone a large piece, passing plates along in the silence.

While Vasilis finished his dinner, shouting to his wife to bring him another beer, Michaelis sat glumly at the next table chewing his nails. Vasilis shovelled in the macaroni and cheese, his loud chomping sounding in the empty room. When a hippie couple stepped in, Vasilis ignored them.

Maria harangued Michaelis because he would not eat. Exasperated, she set in front of him a heaping amount of macaroni and cheese and ordered him to eat it. Michaelis shouted to leave him alone, the first words he'd spoken. Vasilis joined his wife in her attack, advising her to fetch him a beer. When Maria took salad to the hippies she also set salad down on Michaelis's table, by now a complete dinner served him against his will.

Michaelis watched Vasilis's plate. When Vasilis had finished he stood up and tapped the table with his packet of cigarettes. He slouched out of the door into the dark. Vasilis followed him. When Maria came out from the kitchen and saw her uneaten food she shouted after Michaelis that he was a stupid boy.

'So?' asked Vasilis. The two of them faced each other where the pavement gave way at the edge of the village. 'What is it? You come into my café. You don't speak. You refuse my wife's food.' Vasilis wanted to hear how Michaelis would explain himself. Something had changed the boy's mind because that afternoon when Vasilis found Michaelis at the back of the café using his hose and suggested while Michaelis washed himself that it was the night to dig for treasure, before the archaeologists got it all, Michaelis refused. He wasn't interested. He told Vasilis to find someone else. He was off, he said, on an excursion to the mountains. He'd be back too late. Vasilis argued but Michaelis was adamant.

'Why have you come back from your excursion so early?' Vasilis teased. He enjoyed the boy's dismay.

'I didn't go.'

'Why didn't you?'

'I changed my mind.'

Vasilis stopped there. He would not hurt the boy's pride. Calculating the gain if Michaelis were to do the job, Vasilis now complimented him on his good sense.

Finally Michaelis broke out of his mood. 'The drunkard Georgos is guarding the tomb tonight because the old man's gone home,' he began. 'I was at the café just now when the police came in with the news about his son.'

'What news?'

'His son's in hospital. He was attacked.' Michaelis slapped the back of his head. Vasilis waited for Michaelis to go on. 'Georgos is a drunkard. He's sure to be asleep by now. Lefteris took him raki. He'll have drunk it all. He can't stop himself,' Michaelis warmed to his subject; his courage grew in the safety of the empty street. His need for revenge pushed him on.

'How do you know the old man's not there?'

'He asked the director to let him go home. He's a broken man. He thinks his son is dead.'

'Take a bottle in case the drunkard's still awake and pretend you're visiting him. Talk until he falls asleep,' Vasilis ordered. Michaelis nodded, humbled by such cunning. And willing. Perhaps now Vasilis would live up to his reputation. The night after the storm had been Michaelis's initiation into the business. He and Vasilis had visited the tombs several times. But it was the night after the storm that they'd dug through the rest of the first tomb of the English. They found only pieces. Michaelis was disappointed, his appetite whetted by Vasilis's boasts of other triumphs. Those few bases and handles weren't worth it. They'd missed the gold! They hid the sack of handles and bases in Vasilis's house by the old village. No point taking it to his cousin's in Ierapetra. Only whole jugs were worth anything. When Michaelis goaded Vasilis about the gold they'd missed Vasilis blamed it on the wind; the dust had blinded them. 'Tonight we'll find gold!' Vasilis hissed into his ear. It was a dark night, low clouds sealed in the darkness. Nor was there wind; the heavy, wettish air boded a change in the weather. 'We'll beat the English and the rain!' Vasilis laughed and Michaelis laughed, the red tips of their cigarettes revealing them to no one, the two men alone on the coast road, turning back now towards the riverbed where Vasilis had parked his cousin's car. They must wait another hour. Then they would drive the car around the back way up to the old Turkish road at the top of the valley in order to miss the village and approach the hill from the other direction.

Georgos was asleep, curled up under the olive tree, snoring. As Bill and Andonis picked their way up to the site they heard the snore before Bill's torch found the curled-up figure wrapped in a blanket, sound asleep. Bill ran his torch over the trench to make sure it had not been disturbed.

Striking match after match Andonis found the *stamna*, poured water into a bucket and threw it on Georgos who yowled. Bill cursed him and shone the torch into Georgos's face. There was an empty bottle beside him; the dead embers of a fire showed how long he'd been asleep. The dishevelled Georgos groaned and stretched, and slowly got to his feet while Andonis blew on the embers to nurse a flame back to life, slowly adding the smallest twigs until there was again a fire.

Andonis found himself a stone and sat down. He zipped up his jacket and pushed his hands into the pockets. Bill wished him luck. Andonis didn't answer; he stared in dudgeon into the fire. Bill had already gone some way when he returned, anxious to have some response from his workman before he went to bed. 'Your son will be all right,' Bill called. 'Don't worry. Go tomorrow. There's the afternoon bus. All right?' No answer came and Bill gave up, cursing the childishness of the Cretan.

Michaelis found Andonis wide awake. The upright figure leaning forward on his knees by the fire was a shock. Michaelis switched off his torch and waited, his heart beating wildly. Why hadn't Andonis gone home? He'd been certain it would be Georgos still and that he'd be so sound asleep that it would all be easy. He'd planned not to wake him up. But he was awake, and it wasn't Georgos anyway. It was the old man Andonis who'd worked for the English all his life and was their friend. He'd even not gone home to his son that night, he was so loyal and devoted.

As quietly as he could Michaelis crept away back down the hill, meeting Vasilis half way. He told Vasilis who was up there wide awake. They must give it up.

But Vasilis wanted the gold. He did not want to give it up. This time they'd find everything there was. A whole treasure, a king's tomb, gold, precious stones, ornaments, idols were waiting just there, at the top of that hill, to go into his sack. Vasilis held up the sack he'd been clutching under his arm and shook it at Michaelis, calling him a coward if he wouldn't come with him and help.

'But what about Andonis?' Michaelis whispered hoarsely. 'What do we do about him?'

'Go,' Vasilis ordered him, 'and ask him about his son. I'll wait here. Leave your pick here. Go as a friend. Talk to him.'

Michaelis stumbled back up the hill and waved when he reached the top, a wave Andonis couldn't see. But he called to Andonis and Andonis answered, asking who it was.

Vasilis's greed inspired Michaelis. His courage was coming back, and his cunning. It came to Michaelis now that Andonis was after all a father and he'd seen him only a few hours earlier weeping, distraught over his son. 'Why are you here?' Michaelis asked Andonis now, walking boldly up to the fire. 'Why haven't you gone home?'

A long moan came from the old man. He shook his head. 'My son,' he said. 'He is my life, my hope, my pride. He is the world to me. Everything! And all he said was to move along. They hit him on the head.'

'Shouldn't you be with him? Go to him. Go now.'

'To walk home would take days!' wailed Andonis. 'I have only two legs and two feet. They are too slow.'

'Wouldn't the director drive you?' Michaelis asked, squatting down in front of the fire.

'That man has no heart,' Andonis exclaimed.

'Vasilis's here with his cousin's car. He would take you part of the way. Ask him.'

'Why is he here? Has someone sent you? Why are you here?' Andonis stood up and peered into the night but could make nothing out beyond the orange light of the fire. 'I don't see him. Have you brought news of my son?'

Michaelis whistled to Vasilis and told Andonis to sit down again and listen. The man was in agony. It was easy now for Michaelis to see a way to persuade Andonis to help them. It would be for his son that Andonis would punish the director. It would be a just revenge, and afterwards, Michaelis promised, they would take him to his son.

The three men worked long into the night, Vasilis fetched more sacks from the car, Andonis shovelled, Michaelis picked with quick, deft strokes, careful not to break anything that was whole.

By dawn they'd gone. The hill slowly grew out of the night as grey light sifted through the darkness. The two upended wheelbarrows leaned against the tree, their metal legs jutting out into the dawn. The picks, shovels and ranging rods were close by just as they'd been left, the only change a mountain of dry earth by the side of the unrobbed tomb, the trench now cleared down to bedrock.

CHAPTER 20

Laura's bad night was Christopher's fault, she'd decided; in the middle of the night she blamed her mix-up with Michaelis on him. If Christopher had only involved her more in the work and not left her to sweep and fetch and sieve like a fool while everyone else did the skilled jobs of picking, drawing and recording, she'd not have needed Michaelis. She'd have had the confidence to say no. But Christopher had made her feel like a nobody. Yesterday Michaelis had hidden behind the rubble heap, confident she'd go off to the mountains with him. She'd misled that poor man, using him to get over Christopher's neglect.

'I'm not stupid,' Laura cried, on the attack, as they climbed the hill in the grey morning light. Christopher's serviceable brogues made their relentless ascent like tractor wheels up the rocky slope. Laura shouted at him to stop a minute. He stopped and turned. With her leg propped on a firm rock, her right hand gripping her hip, Laura stared up at the tottering figure of Christopher, who was blatantly put out by her harangue. 'Perhaps this isn't the moment,' she admitted, 'but I can't stand any more. You treat me like an idiot. I'm a quite intelligent person. I got my second in English, and I'm not willing to be ordered about by you!'

'I'm sorry,' he apologised.

It was the way Michaelis had jumped out at her, grabbing her arm as if he owned her, that was such a shock. 'I don't want you to apologise,' she cried. 'I want your respect.'

Christopher gave a doleful shrug and continued up the hill, once again leaving her to feel foolish as she followed behind. Christopher was like a blinkered helmet in a museum, his armorial uncouthness an English anachronism as it lumbered up this barren hill on the south coast of Crete at a quarter to six in the morning. He was the most infuriating man she'd ever had the misfortune to meet. No wonder England was going to the dogs. With Christopher Bendicks at the helm it was destined to sink in its own crass arrogance.

Laura ran after him. She scratched herself and bruised an ankle. On the brow of the hill she caught up. 'At least try me,' she beseeched, at the very moment that they reached the top and saw the heap of earth.

Christopher ran, and tripped on a root. Laura thought he'd fall. He teetered on one foot like a stork while he rubbed his ankle. Susan and Edward caught up with Laura, panting from the climb. Susan started to ask where Andonis was. She stopped in mid-sentence. Christopher stood paralysed. Susan made a wide circle around to the other side of the trench and stared into the empty hole.

'Oh darling, how awful!' Laura put an arm around Christopher who looked woebegone, his hands hanging lifelessly at his sides as he peered into the hole. He looked so hurt. Susan and Edward averted their eyes as if it were Christopher's fault, his shame unsightly. 'What bastards!' Laura exclaimed. She meant Susan and Edward, although it was assumed she meant the robbers. 'Thank God we got all that stuff out yesterday. Wasn't that lucky!'

Manolis and Georgos left the English to themselves with a tact that Laura thought wonderful. She watched them, with their backs turned, light cigarettes. They were silent. They walked to the carob tree on the far side of the hill and crouched down against its trunk.

'Shall I set to and start sieving?' Laura ran her fingers through the heap of dirt. Her concern for Christopher made her frown. She was full of chagrin that she'd harangued him only minutes earlier. The least she could do was suggest a menial task for herself.

'Sieving.' The word caught Christopher's attention. He'd felt Laura's hand on his shoulder. And now, how amazing, she suggested sieving. She was right. They should start sieving immediately.

Christopher lifted his head. Susan and Edward were whispering together. Laura wondered what he'd say to that censorious pair. 'What do you think?' he shouted across the empty trench to where they were crouched.

'It's obvious,' Susan retorted. 'Isn't it?'

Christopher looked for Laura. He found her a few feet away holding the sieve; she waited patiently for him to give orders. He smiled. She saw the tears. Of gratitude? He found a comfortable place for her to settle, clearing the spot of stones with his heavy shoe. 'On with the sieving,' he announced. He ignored the snigger that tilted the line of Susan's small, thin lips. 'And you, Susan, fetch Georgos and Manolis and get on with A5 and 6 trenches,' he ordered her. Edward scrambled to his feet and followed Susan. Christopher's anger frightened them.

Christopher didn't speak. Laura peeked up at him and saw how shock had blanched his handsome face, the blond eyebrows in a frail frown that would shield him from questions. He set buckets of the earth in a ring round her. She was near tears herself, so fiercely did she repent her attack on the poor man only minutes earlier. She blushed at the memory of the 'At least try me' shouted at him with such venom at the very moment he had to confront the obscene heap. She would sieve for all she was worth.

As Christopher walked back along the valley, passing a bow-legged woman in dusty black dragging a sack of twigs, he felt sick. His legs moved giddily, no dependable strength in them. Around him everything looked alien and threatening, the dark green promise of the orange groves on his left immaterial, however rich and consoling for the village. There was no consolation anywhere for him now; this country he knew so well was a farce that ridiculed his trust and exposed his vanity in believing he was loved.

Andonis gone! The fire dead. His absence shocked Christopher more than the empty tomb. Was Andonis a robber? In a fury Christopher blamed Bill. Bill had not let Andonis go to his son. It was cruel and stupid of Bill. It was Bill in a temper, when he did wrong because he was furious. He'd forced Andonis to betray his friends in order to have his revenge.

What a muddle! Punishment and revenge were dreadful. It made everything so difficult and sad. He loved Andonis. He hated to think of Andonis in prison. He had a jolly wife who had often given Christopher meals, and two beautiful daughters. They'd all suffer the disgrace. And starvation. They needed Andonis's wages. How would they live? Spend all they'd saved for the daughters' dowries. Christopher moaned. He pressed his hands to his temples. No, no, no, he mustn't let it happen. He must save Andonis. He was sure if Andonis had done it, he hadn't meant to. The police were the problem. What was he to tell them? Could he say that Andonis had returned home to see his son in hospital? But the police would want to know who was the night watchman. Andonis was the night watchman and by morning he'd gone. But perhaps he went before the robbers arrived, too anxious to remain all night away from his son? Perhaps that was the truth! Before reporting it to the police, he must try to find out if Andonis was on his way home.

Bill was standing at the top of the staircase when Christopher reached the Kallithea site. He and Badger were laughing, as Mary Elizabeth brushed clean a fifth step. But when Bill saw Christopher weaving his way between the piles of rock toward his old trench Bill threw down his cigarette.

'Where's Andonis?' Christopher called, shading his eyes with his hand. 'He's gone.'

Christopher sat down on a pile of rocks and wiped his forehead; he felt faint from his long walk in the heat. It was a muggy day, the sticky grey air was suffocating. 'Hot,' he mumbled, staring dumbly into the distance. 'How's everything going?' he asked.

Bill stood over Christopher with narrowed eyes. He waited for Christopher, who sat huddled like a guilty child, to explain what was wrong. Badger watched from a few yards away.

Badger expected to be summoned. He tried to guess what it was. *O Klystopha* was waving his arms. It was Bill the director who was the calmer of the two as they withdrew to the edge of the hill which overlooked the sea. There they stood, the director's head tilted down in a listening attitude, *o Klystopha* uncharacteristically gesticulating with clenched fists. Badger was too curious to keep away. Poor *Klystopha*, whom he liked, looked ill. Badger touched Christopher's arm and asked if he was all right. Christopher jumped because he hadn't seen him come. His damp face swivelled round. Badger muttered apologies and stepped back. Bill told the foreman what had happened.

'But I don't think it was Andonis,' Christopher interrupted. 'If only I knew where he was!'

Badger grunted. Christopher pleaded with him. 'They'll accuse him. But even if he did it, it wasn't his fault.'

'What the hell is that supposed to mean?' Bill broke in.

'You should have let him go!' Christopher pointed his finger. 'All this wouldn't have happened.'

The foreman looked from one man to the other, keeping a modest distance. They now spoke Greek so that Badger could understand, but he would not give an opinion which might be held against him later. A robbery was serious. Trial and imprisonment were the punishments if the robber were caught. He had his family to consider. The foreman demurred when Christopher beseeched him to agree. He lit a cigarette to avoid answering.

'I think it is fair to say that the tomb would not have been robbed if Andonis had kept quiet about it, which is why I did not let him go off to his son last night,' Bill defended himself.

'To punish him! Which didn't pay, did it?' Christopher argued. He denounced his colleague, which Badger thought strange but justified.

'If you'd been in my shoes you would have done the same.'

'That is where you are completely wrong,' Christopher disagreed. 'I thought last night you were wrong. And I still think so.' Christopher was angry.

Badger smiled, both embarrassed and amused. Even angry, the English remained so English. Bill and Christopher both spoke as if they were addressing a formal meeting. They larded their hate with English standoffishness. They spoke in low voices with fists clenched out of sight, in the privacy of their pockets. Not in the face of the other man, for all to see.

When they reached the police station Bill volunteered to telephone the Ephor. Christopher's relief was a picture. He leaned against the door frame and wiped his forehead with a handkerchief, bowing his head. Did he now regret his aspersions? Christopher had refused to understand Bill's treatment of Andonis as they'd inspected the robbed tomb together before coming to the station. Not once would Christopher admit how maddening it was of Andonis to talk big at the café about the gold. He had given Bill no leeway to hate himself less for the whole horrible mess. It was hard on Bill for the very thing to happen which he had anticipated would happen if they weren't careful! But Christopher showed no sympathy. Christopher's only worry was Andonis, convinced that the innocent Andonis might be blamed for the robbery if they didn't do something to protect him. To tell the police lies if necessary was Christopher's high moral position.

When Bill had got the museum's number and was eventually put through to Koraes he pushed the receiver into Christopher's hand. The trick worked. Christopher stared at the black receiver as if it would bite him. He had thought Bill had let him off. Bill moved round to the other side of the long table. All eyes were on Christopher, who held the telephone with the limp, blank look of a silly old man. He started to shiver. Bill looked away. He heard the halting phrases that eventually came out of his fellow director with contempt; he noticed that Christopher's arms were now folded tightly in front of him as he reported to the Ephor.

On their way out it transpired that a policeman had parked his motorbike across their path, forcing Bill and Christopher to climb over a rusty oil drum to reach the street. Bill who was in the lead did not look up, so did not recognise the man who'd waved a gun at him in Ierapetra. It seemed just another instance of a Greek policeman showing no consideration of others, not anything to be remarked on. Bill brushed off his trousers when he reached the road. Christopher followed, bending forward to get over the drum; Bill took in the leathery wrinkles on Christopher's neck and a blue vein he'd never noticed before that twisted up the inside of his arm. Such an intimate glimpse of his friend's disintegration disturbed Bill. He started to walk on, dreading the moment when Christopher would catch up and tell him what the Ephor had said.

It was the last straw to find Jenny weeping. Christopher stood irresolutely by the door. He and Bill hadn't spoken because Christopher had lagged behind. He needed a drink but was stopped by Jenny's tears. Bill was already seated at the table, and looked put out by his wife. He cleared his throat.

'What did the Ephor say?' he asked, stubbing out his cigarette.

Christopher sat down. He pulled out his flask and swigged down the brandy, shutting his eyes, but he couldn't make the vision of Bill go away. Bill amazed him. When Bill had offered to telephone he thought Bill meant to be kind. He was surprised. And delighted that Bill would suddenly volunteer to undergo the task of telling Koraes. He had thought Bill was showing how much he wished to end the quarrel and be friends. Which Christopher welcomed. And he'd given Christopher time to be grateful before he thrust the telephone at him. The viciousness was breathtaking. Bill blamed the robbery on him, he supposed. Christopher opened his eyes. Why on earth was Jenny weeping?

'The Epimelete's coming tomorrow,' Christopher informed Bill.

Jenny blew her nose.

'What else?' asked Bill.

Slap! Jenny swatted a fly. Bill grabbed the fly swatter. He threw it, hitting Christopher by mistake. Margaret came through the door at that moment and picked it up, taking it with her into the kitchen. Christopher leaned his chair against the wall, aware that Bill hadn't meant to hit him. It stung where the wire scratched his arm but he ignored it.

'Hell,' Bill muttered. He pummelled the table in front of Jenny's place: 'Damn your bloody flies!' Bill stuffed his hands in his pockets and tried to recover a calm that befitted the director. He frowned at the table. Christopher burst out laughing.

'What's wrong?' Bill snarled, which delighted Christopher. He felt light-headed, his sanity rescued. To hell with the bloody Ephor, he thought pleasantly to himself, offering Bill a nip of brandy which Bill refused. Encouraged by the sound of sizzling and smell of onion and garlic and olive oil drifting through the partition, Christopher decided he was hungry. He would not move until he'd eaten.

Jenny moved up and down the table setting out glasses and cutlery, pushing in chairs that were in her way. It was as she was coming through with the jugs of wine and water that she told Bill and Christopher that she'd seen Andonis in Ierapetra. He was in his white shirt waiting to climb up into the bus. She'd honked and waved. She even leaned her head out the window and called, but he wouldn't look around.

An hour later when lunch was over and everyone had gone to fetch their passports, Christopher still sat, his eyes a glassy blue by now, staring blindly at the wall. He was too drunk to move. He laid his head down on the table and fell asleep.

CHAPTER 21

Jenny had been shocked by the sight of the old man in tears. Andonis's mouth gaped in grief, and a babyish whine broke from him as he stood in the doorway. A man usually so dignified! It was so pathetic that Jenny had wanted to hide her face. But when Bill took Andonis off with him to be night watchman the incident left her mind. Other more persistent memories pushed it aside; it had not been frightening like Adam's quarrel with the policeman and the policeman's pursuit of them on Saturday night. This plagued Jenny all Sunday while Adam was with Bill inspecting Christopher's new trenches. Later Adam found her on the beach to talk about colours, and how shadow defined the shape of a pebble; he held up pebble after pebble to show her what he meant, presuming his pebbles interested her. What if the policeman were to hurt the boy? Or her? He had a gun and was violent. Adam had insulted him. Weren't they in danger? Jenny had pretended to listen to Adam about the colours in grey and the colours in green and the magic of a well-drawn line. She thought how sweet and naive he was, and how stupid to have had to have the last word on Saturday night.

On Monday Jenny had avoided Adam, which was easy since everyone was busy and excited about the new tomb. But that night she woke up in her musty room and lay winded from horror at Adam's impudence. What if the policeman had pulled the trigger? She begged Margaret to go with her on Tuesday morning. She needed protection after her bad night (although she didn't say so), but Margaret was anxious to wash her hair and lie on the beach. There wasn't much to buy, she pointed out, handing Jenny the list.

It had been a relief to spot Andonis on her way in. Of all the workmen Andonis was her favourite. Many times she and Bill had been given huge meals by his kind wife. He was proud, with the sort of pride which was mostly self-respect, but also had a hard edge of arrogance which gave him grace in the way he walked and danced and talked. Jenny had always admired how he held his shoulders right back; when he spoke his words came out clearly and precisely, not in gusts of gruff patter as with the others. His other son, not the one who'd been hit on the head, was at university. Andonis hoped he would become an archaeologist. He wasn't set as the other workmen were on making money. He had vision – a prince of a man of whom Jenny passionately approved.

Unnerved by her sleepless night, she was thrown when he wouldn't wave back. He refused to recognise her. Angrily she drove on and parked the car in the narrow street behind the market.

She had not only food to buy. Adam had asked her to buy him poster paints, and had given her a list of colours. She was afraid she wouldn't find the right shop and was anxious to start her search, resenting that the food shopping must hold her up. Most of all she was eager to get out of the town before the policeman saw her.

Jostled by other shoppers Jenny tried to keep calm, gritting her teeth not to blurt out abuse in her bad Greek when the sharp prickles of a woman's basket poked into her thigh. In front of her was a stall of tomatoes and cucumbers; the acrid smell of vegetables closed in on her. She bought three kilos of tomatoes and two of cucumbers blindly, waving at the stall-keeper a crumpled 100-drachma note. She wouldn't panic; nevertheless she felt herself near to it. She swallowed hard to keep a hold. If some of the tomatoes were bad, it didn't matter. She'd let the man choose, too weighed down with pork chops and apples to care if he cheated her.

When she could dump her heavy plastic bags in the back of the car and start off for the centre of town, her spirits lifted, freed of food.

Jenny took long strides when she walked on her own, unconscious of others. Until she tripped on an empty beer can. She toppled forward. Had anyone seen? Her spirits fell. How foreign and preoccupied she must look. Unconsciously she'd crossed the street where the beer can lay, and found now that she was on the post-office side. The policeman. She'd forgotten to keep watch. Was he anywhere? Her fear was back. The post office looked like a police station with its new white walls and regular door. Jenny looked right and left and down before she crossed the next street; she scrutinised the shop fronts, practising how she'd ask for poster paints in Greek. Now every dark doorway could be hiding the policeman. Adam's cheek had made the town unsafe; Jenny understood the desire for revenge. She was proud and respected it in others. That policeman had been hurt and he'd make them pay. He'd waited outside the restaurant. When Adam kissed her, the policeman had despised them. Foreigners who were disrespectful and then undignified deserved the worst. How self-deceiving of Adam to think policemen hated lovers because lovers threatened their authority. That poor thin young man couldn't know. He wished so much to be a handsome, romantic big man like Bill that he couldn't see what the policeman saw. He couldn't see how unseemly it was for a youth to kiss a grey-haired woman in public.

In the car Adam had said they'd pretended to be lovers to annoy the policeman – Laura was his dream of a *real* lover. That hurt. Adam had both hurt her and offended the policeman. Why was she now stepping into a dusty school-supplies shop to find him paints and a sketchbook? Even eager to do it! This need to please the boy had become an obsession.

She needed to please the boy because she so obviously failed; it rankled that he preferred Laura. It was she who all his life had been his friend. Bill was Laura's type. Jenny needed to save Adam from his delusion; it touched her that he was so frail and unattractive. His company comforted Jenny, his idealism excited her (when it wasn't dangerous), his puny body moved her. Those bony, sunburned arms looked so untended. She must do her best to find the paints.

Inside the shop the woman behind the counter produced a dusty box of tubes. Jenny smoothed out the crumpled envelope on which Adam had scribbled 'alizarin crimson, cadmium red, cadmium yellow, lemon yellow, French ultramarine, Prussian blue'. She was exhilarated by the professional sound of the colours.

But she must have looked worried as she fell on the tubes with the envelope in her other hand. The bosomy shopkeeper in a cool-looking pink dress gently pushed Jenny back and turned out the contents of the box on to the counter, spreading out the tubes with quick, deft hands so that Jenny could see more easily what there was. Jenny blushed. She caught the woman's eye and gave her a meek smile. The kind woman was interested. She asked Jenny to explain what the words on the envelope meant. Jenny launched more bravely into Greek, encouraged by this kind woman in a pink dress in a shop smelling of glue and new books, ink and plastic pens. Together they found the colours Adam wanted (which he'd hinted Jenny wouldn't find in a 'dump' like Ierapetra). Jenny exclaimed, 'Good, good,' with huge relief; the other woman nodded, returning the rest to the box with emphatic efficiency.

This nice woman pitied her. She waited patiently while Jenny groped in her bag for another twenty. Jenny felt the woman's kind eyes on her. Did she look as if there was something wrong? Was it fright? Oh, it was that damn policeman! Again she'd forgotten. Now she remembered. She rushed to the door, her arm still in her bag. Was that visored cap and black moustache in the street anywhere?

She must go. The pork would be spoiling in the hot car. She shouldn't have been so long buying paints for Adam. She scooped up her package in too much of a hurry and dropped the paints on the floor, her handbag still not shut. The shopkeeper came around to pick up Jenny's parcel. She kissed Jenny on both cheeks, pressed Jenny's arm around the package and patted her on the back. Jenny spluttered and nodded, bumped into the corner of the counter and almost dropped the package again when she waved from the door. No one had been so kind for ages. As she hurried back to the car, the dark faces she passed looked more human, the town no longer menacing. She'd spoken Greek well. She was pleased with herself. She'd go and see what they were doing up on the site. Adam would be surprised that she'd managed to find everything.

She stopped to fill up the car at the new petrol station. The policeman was bending over his motorbike at the other pump but Jenny didn't think he saw her. A new motorbike would monopolise his attention. Jenny kept her eyes right down to remain invisible.

He did see her. He recognised the grey-haired foreign woman in the blue English car. He released the trigger on the petrol pump so as not to finish filling up his bike while she was there. He was pleased she didn't recognise him. If Jenny had looked up she would have seen that the policeman smiled. He was all the more pleased because his new motorbike meant he could follow the blue car and find out where they lived. Would she realise? He could keep well behind. He needed to know for his job. They were the sort of foreigners that took advantage of his country and shouldn't be there. He was no longer smiling, sombred by righteous anger. He pulled the trigger and finished quickly. While he paid his cousin who had got the permit and loan to build the station the year before, his eyes were on the blue car. He was to have lunch with his cousin as he usually did on a Tuesday. He waved and jumped heavily on to the pedal of his bike.

'It's very important. I can't tell you now,' Sotiris Georgakis shouted back when his cousin ran after him.

Sotiris Georgakis knew the village, the only big village on that coast with a beach and many houses. When he saw the foreign woman park her blue car in the square he was satisfied and accelerated round the corner.

Two more of the foreigners stepped out of the police station when Georgakis squeaked to a stop outside. The tall dark-haired one he recognised. The other one was a typical Englishman with blond hair and a proud, shoulders-back way of walking. Georgakis loathed the two men instantly. He backed his bike so that it was in front of the steps, forcing Bill and Christopher to climb over an empty oil drum to reach the street while he watched.

His friends inside were surprised to see him, but were too busy to say much, both of them leafing through documents, the officer sitting, the young Lefteris leaning on the table when Sotiris walked into the room.

'Aren't you going to speak to me?' He slapped the table.

The bald officer with Bill's passport held out in front of him looked up. 'What do you want?'

Georgakis fixed his eyes on the face of the officer. He slapped his leg and pointed at the empty doorway. 'I know him. Who is he?'

'Who do you know?' The officer tilted back his chair. 'Can't you see I'm busy?'

'I followed the English woman in the blue car.'

'Fancy her, do you?' Lefteris let out a loud guffaw. 'With the grey hair and ugly legs? What taste you have, my boy.'

The officer told Lefteris to be quiet. Grateful to the officer, Georgakis sidled up to the officer's chair and reached for the other passport on the table. Jenny's familiar face started out at him. Excited, he waved it at the officer. 'That's her. I saw her kissing another one of them.'

'Where?' asked Lefteris.

'In the middle of the street.'

The officer shrugged his shoulders. He showed Georgakis Bill's passport. 'I suppose you mean him.'

Georgakis squinted at the photograph of Bill in an expensive-looking sweater. The same proud eyes which had tried to ignore him outside the post office looked at him, but then he had been wearing a striped shirt only, with sleeves rolled up. Here in the photograph he was dressed well because he was home in England where he belonged.

'Well?' The officer yanked back the passport, irritated by Georgakis's obvious short-sightedness.

'No,' Georgakis shook his head.

'No what, my boy?' asked the officer.

'She wasn't kissing him. She was kissing a thin youth.' Georgakis pursed his lips. Lefteris and the officer exchanged looks; Lefteris pulled up a chair and sprawled out in it. He pulled out his worry beads which made a light tick-tick sound, as Georgakis held out his arms to mock an embrace.

'I've seen *him*, too, with a beautiful young brunette.' Georgakis jabbed at the photograph of Bill. 'Small breasts with nipples like cherries!'

The officer grunted; Lefteris ran his tongue across his upper lip, giving the beads in his right hand a jerk. Georgakis rolled his eyes.

'I know the one,' Lefteris mumbled; a throaty laugh revealed he'd noticed those nipples like cherries. His jacket hung open as before; tufts of wiry black curls spilled over his white vest. Both he and the officer were unshaven, only Georgakis was shaven and neatly buttoned up. His cap on the table beside the passports gave his visit an official look.

'You should stay,' advised the officer. With his stubby fingers he made a neat pile of the passports. 'You know something. I'll need you to find out where they all were last night.'

Georgakis looked from one to the other, thrilled that they were to take him into their confidence. He held his breath and raised himself on tiptoe, the moment holding for him such promise of joy that he thought he'd burst.

'Tell me what's happened,' he whispered to hide his excitement, his eyes raised up to the ceiling in what he hoped was a bored, off-hand manner.

Jenny drove the car into the sliver of shadow on the east side of the village square. She avoided the white stone monument, wary of its new, hot orderliness in the middle of the messy square. It had a vindictive air about it, so clean in the midst of the old sacks left out from the warehouse and the weathered, broken doors of the low houses which looked blind to this gleaming memorial to the dead. It took the Colonels to put it up years after the poor old widows had lost their husbands, their mourning long since turned to dust. Was it to whip up the old outrage that the Colonels scattered such brash monuments all over Crete? Did they think they were avenging the injustices of thirty years ago? Jenny was in jitters from her drive. She despised harking back. Her mother's pet subject, Jenny bemoaned as she hurried back to headquarters, was how the cruel and grasping English had burned down the White House in 1812. She and the Greek Colonels were of a piece!

It had been the visored cap every now and then bobbing into her rear-view mirror that eventually caught her attention. That horrible policeman had followed her. Why? What had she done wrong? It frightened her not to know. Anger at Adam for creating a situation which might have gone away if they'd avoided the man now made Jenny's head ache and her mouth dry as she washed the pork and sprinkled it with lemon juice. Margaret had not appeared to tell her what else needed doing. When Bill burst into the front room and shouted for her she responded in a weak, tired voice.

'Damn it, where've you been?' Bill loomed in the kitchen doorway, pushing his hand through a wild confusion of curls as he accosted his wife. 'I've been looking everywhere.'

'I'm sorry.' Jenny kept her eyes on the pork. 'I've just got back.'

'Why'd it take you so long?'

Bill's reproach was unfair. Hunted down by a policeman, she still tried to do her duty although sick from fright. For her husband to come in and blame her for taking too long stung her. She threw the lemon at him hard. It hit the door frame and dropped to the floor. Jenny burst into tears. 'Shut up,' she wailed, sniffing childishly, wiping her nose with the back of her hand. Guilt that it had not been the tomatoes and pork for Margaret which took the time spurred Jenny now really to weep. She collapsed into a chair and covered her face with her lemony hands.

Neither Christopher nor Bill could think what to say.

If Laura had gone straight to her room after lunch Michaelis wouldn't have pushed open the door into headquarters and found Christopher asleep in his chair. Christopher was alone, his arms stretched across the table, his cheek pressed against the tabletop in an uncomfortable way. But before Michaelis could back away, Christopher woke up. The juddering of the door woke him. The scrape and creak reverberated between the two men. Michaelis stood awkwardly in the doorway. He apologised. He could not admit that he was looking for Laura. His eye had already taken in her absence.

'Where's the director?' he asked, moving sideways into the room so that he could see Christopher better. 'I was looking for him.'

Christopher pushed his hair back, showing by the way he smiled and waved to Michaelis to sit down that he suspected him of nothing. Even this lie that he was looking for the director had gone undetected. What could Christopher think he would want with the director? Michaelis shrugged and pulled out a chair. Did Christopher think he had information? Was that how to continue with Christopher, who still looked groggy? He offered Christopher a cigarette which he accepted.

'I've never seen you smoke before.' Michaelis inhaled deeply. 'The director smokes a lot. Not good for him. Not good for me either!' Michaelis pressed his hand to his chest. 'I'm doomed.'

Christopher puffed at his cigarette and then examined it as if it were the first time he'd laid eyes on one.

'Are you afraid of death?' It was not the question Christopher had expected. Michaelis smiled. 'I am,' he carried on, not waiting for Christopher to answer.

It was caution that kept Michaelis in his chair at the end of the table, afraid that if he left too abruptly Christopher might wonder about him. His smoking was hectic as he turned over in his mind what his next step should be.

'I have information about the robbery,' he blurted out rashly, feeling more and more uncomfortable. It was Christopher's silence now that unnerved him. If he'd only say how upset he was or who he thought had robbed the tomb! He must already suspect Vasilis. Everyone else in the village did. But Christopher studied his cigarette with that faraway look so typical of the English lord type.

Michaelis cursed himself for coming to this 'headquarters'. It was Laura's fault. Did she regret now not going up to the mountains with him? That was what he needed to know. Did she realise she'd caused the robbery? They'd have had their cups and plates and figurines. There was a mountain of stuff that Vasilis hid in his house by the old village until his cousin could collect it. Vasilis needed the money. He was in debt. And greedy. He even hated his cousin because he had a pretty wife and two cars. Vasilis not only wanted to be the richest man in the village but the richest man in Ierapetra. He was a fool.

Michaelis didn't care about the money Vasilis promised him. He was in love. The most beautiful woman he had ever seen had transformed his life. Laura was a miracle. Vasilis's plan was that Michaelis use Laura to find out what they were digging up. If Vasilis had a heart he would have seen the danger. Laura was no ordinary woman. She was a goddess. She had inspired Michaelis to make love in a way no woman had ever done before. Michaelis shut his eyes, pretending to Christopher he was tired. The memory of their lovemaking on Sunday night incited such longing in him that he was afraid Christopher would see. They'd gone back to the orange groves. She'd lain on the ground with the graceful amusement of a faun, her long legs folded under her, her thin shoulders bent forward, her long arms stretched up to him as if he were a god. Her long, agile fingers had tingled down his back with an intensity that scattered every scruple; she'd opened her wide full mouth to him, smiling. Smiling! For hours they'd lain there in each other's arms, her passion another world, far far from his life at home with his mother and father and his three unmarried sisters, getting what work he could and sitting long hours at the café with his friends when there was nothing else to do.

But she'd refused to come to the mountains. He needed her. He was delirious. While he robbed the tomb he was hating Laura for rejecting his great love. Now, though, that had gone. He'd watched her all morning sieve the pile of earth and was crushed. It pained him to see her struggle with the weight. Now he strained to make amends. Did she suspect him, or did she think it was the old man Andonis? He needed to hear her tell him she was sorry.

He had waited behind the rubble heap, but Laura never came. In order not to be seen he walked around past the church and the school. Just as he was passing the school the schoolmaster came out. He'd heard from Vasilis that he was mad on the subject of Germans. He had no heart left, according to Vasilis who was quite heartless himself, but not from hate. Michaelis ducked his head, conscious of his red hair.

'Stop. Come here,' the schoolmaster ordered. Michaelis stopped. 'Who are you? You work for the English archaeologists, don't you?'

Michaelis nodded.

'Where are you from?'

'Ierapetra,' he answered the schoolmaster.

'I know who robbed the tomb. Come here. I'll tell you.'

Reluctantly Michaelis had moved nearer the short man with his cane, who looked very clean in a white shirt. That Michaelis's arms were still dusty from digging made him ashamed; he rubbed down the front of his brown T-shirt.

'The German. Ask him. I know. That's why he's come to this village. You tell the director I said so.'

Michaelis had started to move on when the schoolmaster shouted after him that he should not stay long in the village because Germans, even bastard sons, were not welcome. If he hadn't already moved too far away Michaelis would have knocked the man down. 'I'm as German as you are,' he shouted back. Then he broke into a run, ending up here at headquarters, hoping to find Laura.

But there was only Christopher – alone, asleep – whom he'd unfortunately woken up when he pushed open the door. No Laura anywhere. Only Christopher waiting to hear what Michaelis had come to tell the director.

'The robbery has upset us very much,' Christopher suddenly said, staring at his cigarette. 'I hope it was not someone from this village.'

'Where is Andonis Markakis?' Michaelis asked disingenuously. The wary look Christopher gave Michaelis in answer unnerved the redhead. Michaelis rubbed his eyes, adding quickly that he supposed Andonis had gone home to his son in hospital. He lit another cigarette.

'The schoolmaster thinks I did it because I'm German!' He pulled at his hair.

'That's ridiculous,' Christopher exclaimed, refusing Michaelis's offer of another cigarette.

'Then come and have a brandy,' Michaelis commanded, jumping to his feet with fresh energy.

Laura couldn't stop the tears rolling down her face, it was so painful. She lay on the beach waiting until the pain let up enough for her to stand up. There were only two hippies in bras and jeans, with their backs to her, 200 yards away, sucking at pomegranates. She gritted her teeth to keep back the groans. But the icy pain was so intense that it made her feel sick. Eventually it slackened and she could sit up; she wound her scarf around the same weak ankle she always turned when she was low. It was years ago when she'd first turned it playing tennis. She'd insisted on finishing the game, thinking she was wonderful to hobble after the ball. Her ankle never recovered. Every time her life was in a muddle, it alerted her in this painful way as a punishment. Oh, how she regretted her vanity all those years ago when she should have gone to a doctor at once and had her ankle looked after. Her vanity must pay, because every time she sprained it the ankle was thicker when the swelling went.

Laura pulled the scarf tight and squinted up at a strange grey cloud overhead. The sun behind it sharpened its dark edge against a colourless sky which made it a portentous thing. When Susan had said at lunch that it would rain Laura had laughed; she thought Susan sounded too fuddy-duddy for words, like a nanny who warned her charges that their fun would soon be over, so *not to complain and enjoy themselves and be good*. She'd offended Susan and looking now at the ominous cloud and slate-coloured sea rolling cold waves onto the black shingle, she suspected Susan had been right. The pain had receded to her ankle, leaving the rest of her detached and frail. It was on her honeymoon that she'd turned her ankle the second time, when Francis left her groaning in a ditch while he went to look at the ruins of a Romanesque church. He came back half an hour later full of smiles to see how she was but when he started to describe the beautiful carving, she called him cruel. He was dumbfounded. Francis thought leaving her until the pain was less was the only thing to do. She cheered up when she had the walking stick and could pretend she was a woman of great age as she scowled at imperfect pastries in the shop windows. But Francis fumed for days, peeved that she should have misunderstood. And here was the ankle again that had ruined her honeymoon – a past muddle. Their lovemaking had not been progressing well, Francis so preoccupied with the itinerary that he had a guidebook propped open on his knees five minutes after they'd at last consummated the marriage.

What was it telling her now? She would not try and answer that. She rubbed her hands slowly up and down both legs and fixed her gaze on the grey sea, to merge her consciousness with the thin, faraway sound of the waves.

When Laura spotted Christopher at a table with Michaelis at the other end of the street, embarrassment made her ankle throb worse than ever, tucked under her in a wounded-fowl way. The kind widow who owned the bakery near the beach had rushed out when she saw Laura through her door. In her black dress covered in flour, with sleeves rolled right up to her armpits, she pulled Laura to her, wrapping a strong arm round Laura's thin waist. Laura leaned more heavily on Kyria Georgia's arm.

'What'd you do?' The moment Christopher saw her he hurried to take Laura's other arm. Georgia told Christopher that she'd seen Laura bite her lip in pain. Georgia bit her lip and moaned to show Christopher. When Christopher followed suit and moaned Laura burst out laughing, bewildering both her helpers. Michaelis held out a chair and Christopher ordered her a brandy. Laura felt near tears again, everyone was so kind. When Christopher offered her his silver flask she apologised. When Michaelis lifted up her leg and set it on another chair she apologised. When Georgia patted her on the shoulder and left to get back to work, Laura apologised in Greek. Gently Michaelis untied the scarf around her ankle and took it inside the café, returning a minute later with it soaking wet. He made neat work of her scarf turned into a cold compress.

Laura closed her eyes and let her arms hang limply at her sides, the brandy a rising warmth inside her. When she opened them Michaelis was leaning over her; his beautiful mouth and wide nose peered down at her. He frowned. Laura touched his cheek. 'I'll be all right,' she whispered, which Michaelis couldn't quite understand. He turned round for Christopher to translate. A small bottle of brandy and a glass waited on the table. Christopher had gone.

'You should have come with me to the mountains.' Michaelis filled the glass and handed it to her. 'This wouldn't have happened.'

Laura gulped down the brandy, disliking the taste.

'Why didn't you come?' he asked.

Laura threw back her head to hide embarrassment. She massaged the muscles in her neck. Her predicament was growing at lightning speed. Could she cope, particularly now with her wretched ankle? She tried to move her toe and pain leapt up her leg. She cursed it, pressing down hard on her knee. Frightened by Michaelis's intensity, which was so different from the self-conscious obliqueness of past English lovers, she wondered what to do.

'I'd love to go some time.'

'With me!'

As if Michaelis were blinding, Laura squinted at him. 'That would be lovely.'

'When?'

'Not with my ankle like this.' Laura poured out more brandy and handed it to Michaelis. He slammed the glass down on the table. Laura jumped. Then he swigged it.

'Do you think I'm German?'

'I thought you were Cretan.'

'I am. I am. But I met the schoolmaster and he thinks I'm German. He thinks I robbed the tomb. Do you?'

Laura was indignant. She could never forget the schoolmaster's outburst before the German plopped the frog on the table. Laura leaned forward, grateful to Michaelis for reminding her of just how much she disapproved of bigots. 'How ridiculous,' she exclaimed with great energy, set on quite another track now. 'Poor you,' she commiserated, patting him on the knee. 'Though what does it matter? Of course you didn't rob the tomb.'

CHAPTER 22

Christopher was wide awake. Outside, a stray length of polythene skittered over the ground, flapping in a panic when the wind nailed it to the door frame. It stopped flapping and flopped to the ground, the messy noise of it grazing Christopher's fury. He pounded his bed childishly as tears streamed down his face. He wanted to throttle that interloping rag of polythene until it couldn't mess up his life any more. Damn everything. Damn everyone. Laura had seemed so sorry for him this morning. She had laid her hand on his shoulder and called the robbers bastards, instantly loyal. All morning she had sieved because she thought he wished it – with a worried frown on her beautiful, dusty face. It was a support to know she had carried on sieving all the time he was away quarrelling with Bill and reporting to the police and the Ephor. When he saw her later at Georgia's side, hurt and in pain, he'd rushed to her . . . fool that he was to believe she'd forgiven him those crass fumblings at Vasilis's. He'd hoped Laura no longer minded. Groundless hopes. Fooled again. She'd raised his hopes so far that he'd even reproached himself for disapproving of her brazen dress and behaviour the night of the backgammon game.

That redheaded Michaelis, of all people! But of course. It had started right under his nose that very same night. Vasilis brought him in. And Laura stayed on. They . . . Christopher squeezed his eyes shut to stop imagining. His filthy imagination was indecent. He turned over on to his side and stared wide-eyed at the wide cuffs of his trousers which hung from the nail, the smart cuffs dangling in front of him . . . She'd touched Michaelis's cheek. She'd whispered, 'I'm all right,' to that damn Cretan when he was standing just behind with the brandy bottle, thinking she might love *him*.

Christopher hugged his pillow, an unsightly wretch. To weep for the exquisitely desirable Laura who had sensibly found herself someone else was so like him. He was actually a very ordinary person who longed to be loved by a beautiful woman, and only criticised women because they rejected him. He was not too old and rich to desire joy. His dirty books were too damn vicarious. He was fed up. If only he could buy himself love. He had plenty of money.

Christopher turned over on to his back again and rested his head on his hands, facing the stained rafters. He had the tomb too to face. He'd bungled that as well. Of course there was his mother and it was all inevitable, the loser that he was. He had good taste in shirts. And his trousers were well cut. That wide cuff on the nail proved it. His mother could not deny that he was a dressy loser. Poor tomb. Even the dead weren't safe from rich fools like himself.

Christopher lay a long time in the close room until darkness unsettled him. The soles of his brogues scraped on the rough cement as he pulled them over. There'd been no voices and creaking of tables next door all this time. Why not? The late afternoon and early evening were Bill's time to check on Ellen's strewing and Jack's drawing. But this evening no one had come down. Too taken up with the robbery, he supposed. The robbers had breached the daily routine, dramatically confusing the English archaeologists. In the next room Christopher tripped on one of Adam's tapes. The sharp corner of a table poked into his thigh because he still wouldn't turn on a light. He groaned and groped for the door. But outside it was just as dark. He cursed the darkness and went back for his torch.

In front of him when he'd climbed up to the road were a couple holding hands. Their dark shoulders touched a few paces ahead. Christopher shrank back, afraid of who it might be. A couple of hippies, he reassured himself. Yet he waited until they were too far ahead to hear him behind. He hunched his shoulders and walked on with head right down. The heads of the couple in front leaned toward each other as the two people, one in a skirt, the other not much taller in long trousers, walked in step; when they turned left into the street of the Merry Widow café and headquarters, they moved apart and dropped hands.

At the turning Christopher stopped. He pulled his hands out of his pockets. With hands on hips he surveyed the clusters of men and chairs, and lit-up doorways. Who robbed his tomb? Someone here in the village? A ring from outside would need a local informer. Andonis had been framed. It was not Andonis. Andonis had known them too long. Whoever framed him didn't know that Andonis loved the English. Christopher clapped himself on the chest, smoothed back his hair which he'd forgotten to comb and hurried on.

At the door of headquarters he caught up with the couple, who turned out to be Adam and Jenny.

Inside were the German in the same pale blue shirt and trousers and his skinny American friend. Bill looked flustered, staring hard at his hands. Jenny walked on through to the kitchen. Adam let Christopher past and retreated.

'He hates me because I'm German. He's telling everyone in the village I robbed the tomb.' The German was pestering Bill for information.

Ellen set the table. Christopher took a glass from her and poured raki from an open bottle at Bill's elbow. He sank into a chair. Laura was in the far corner with her leg propped up, reading a magazine.

'Did your workman, the old man, do it? Andonis?' The breathy way the German pronounced 'Andonis' incensed Christopher.

'No!'

The three men stared at Christopher. Until then Bill had ignored him.

'But of course,' the German butted in, 'you must know because you direct the tomb digging. What you think?'

Christopher felt Laura's eyes on him now as well. 'I don't think anything.'

There was silence.

'Andonis has gone to see his son in hospital and will be back in a few days.' Christopher frowned, but his energy was returning now that he'd tumbled on an obvious way to save Andonis's reputation. He would get word to Andonis to come back quickly. He'd be less suspect. Christopher hoped the Ephor hadn't already had Andonis in for questioning. On the telephone he'd asked Christopher if there'd been a night watchman. Christopher admitted what had happened; he wished now he'd thought to take the blame himself. If either he'd said it had been him and he'd fallen asleep, or if he'd suggested there had not been a night watchman the whole night through because they were all so tired, Andonis would be out of danger. But he'd been too frightened.

'Vasilis is telling everyone Andonis has gone to the States.' The smirk on the American's face made Christopher move. When he tripped on the leg of Laura's chair, he glared at her. In the kitchen Margaret and Jenny were cutting up cabbage when Christopher barged in. Jenny jumped, which Christopher noticed, recalling then that he'd seen her a few minutes ago holding hands with Adam. He winced from embarrassment, and Jenny wondered if he had toothache.

Up and down the dark street Adam paced unnoticed, his hands loosely clasped behind his back, the coolish night-time enveloping him in drama. There was little he could see, the gritty pavement which felt hard and uneven under his thin leather soles was the only sense he had of where he was. But every nerve in him was taut, the doorways he passed cowered from him, his all-powerful importance feeling huge in the darkness. The police station was as far as he went. There the pavement came to an abrupt end and he turned back, keeping his eyes down so as not to stumble . . . It was what Jenny had just told him about herself which thrilled him most. She was afraid and needed his reassurance that the Ierapetra policeman could do nothing to her. He'd told her the threats were harmless; following her on his motorbike was as empty a threat as any other. This had calmed her down. She'd let him take her hand and they'd discussed Picasso. Jenny confessed that Bill called Picasso a doodler, which hurt her feelings. Adam understood why this would be so upsetting and encouraged Jenny to say which period of Picasso's painting she preferred and which of his paintings in particular meant most to her. They discovered how they both hated the 'doodler' school of opinion and despised the crass arrogance of people who thought that way.

'I can never discuss anything with Bill,' she complained, resting her head on his shoulder when he slid his hand around her waist. 'He won't understand. He's a typical academic who twists everything. He won't ever agree, on purpose.'

'Perverse.'

This delighted Jenny and she begged him to go on and explain. The poor thing was desperate to talk. They'd sat down on a rock by the side of the road and watched the silky light drain out of the sky until the sea and the rocks and the billowy clouds merged and they could barely make out each other's faces. He'd never have guessed what a passionate woman Jenny was. So much for his mother's intellectual friends! His mother always said Jenny was too clever and made her feel uncomfortable, but that wasn't what he'd found. They'd kissed. It was the first time he'd kissed a woman so old. She'd listened, too, to everything he'd said about why academics were the way they were. She was surprised he knew. He'd stroked her hair and kissed her on the forehead. When they left their rock and started back holding hands, he'd pulled her to him and they'd kissed more and more frantically, their tongues and lips in a fever, until they had to stop. But her weak sigh at the end was deeply pleasing and he knew from all she'd told him that Bill had never aroused her so.

Adam was happy. Meaning spread over the squat flat roofs, which were darker than the dark sky, and the murky movement inside cafés, with muffled voices of men like the soft wash of the waves. His own elation put everything in its place. He was glad the robbery had happened since if it hadn't he'd never have experienced how robberies affected things. Right now the excitement of catastrophe contributed to his joy; it compacted its secrecy, hiding him and Jenny wonderfully. He would throw himself into the search for the robber with model zeal. The following day a detective was to arrive from the museum in Heraklion. Adam planned to keep an eye on the man, that way discovering as soon as possible who in the village was under suspicion. And in the team.

Could any of them have done it? What about Annabel? She owned an antique shop after all. Adam liked the idea of Annabel turning out to control a tomb robbers' ring, coming to Crete to carry out her plan under Bill's nose. Would the detective suspect her? To clear out a whole tomb in a night was pretty good going. Whoever the robbers were, they were incredibly strong and efficient. Christopher's appearance at the Kallithea site that morning had shown a transformation; until then he had always been neat and brisk and healthy-looking. He was stooped and dishevelled, his straight hair fallen over one eye in a desperate sort of way. For Andonis, Christopher's pet workman, to disappear as he did was almost too suspicious. A frame-up? A frame-up would push Christopher to some sort of brink. Would he fight? Christopher personified for Adam the English appeaser, a repressed coward who avoided confrontation. Had he any fight, Adam wondered. Or had it been educated out of him in his pampered, upper-class upbringing? A distraught, frightened Christopher would be amusing to observe, and gratifying since it justified Adam's impatience with most of his compatriots.

Bill was another English casualty, of a different sort. Adam knew how it infuriated Bill when he mislaid a tape or a ruler, or lost his pencil in the thorn bushes on his way up to the site. It wasn't surprising he dropped things since he had much more to carry than the others, but Bill was too preoccupied and irascible to see that. He couldn't forgive, so frightened was he when anything was lost. Like a child, he needed to blame and criticise, obviously thinking, of Adam, 'what an incompetent wretch', every time. He wasn't incompetent. His drawing of Bill's fortification wall was good: the stones were accurate, a jolly difficult thing to do. When Bill was so angry at lunch because he'd been held up looking for one of the tapes, he'd have hit Bill if he could. Instead he spilled custard on the table and knocked over his wine without apologising; Bill growled as Jenny mopped up the mess. Of course the robbery had Bill on edge. Christopher fell asleep before lunch was over but Bill was jumpy, even calling Mary Elizabeth a silly girl – which was the first time any of them had ever heard Bill criticise the beautiful American – when she put sugar in his coffee.

Bill was to be avoided, for many reasons. Just now seeing him through the door confronted by the horrible German and his American sidekick was delightfully awful, Bill's tousled consternation most satisfactory.

Hunger began to subdue Adam's high spirits. The long walk with Jenny told on him; the street felt harder and more uneven with every step. He stopped at the doorway to headquarters and hid in the shadow of the mulberry tree opposite. Soon the German and the American must leave. The bright dial on his wristwatch told him it was eight thirty, the 'official' hour for dinner. Though he'd often turned up late, which annoyed Bill, he was never unaware of the time. His new wristwatch kept excellent time. The hours of the day had always coloured what he did – to start out on a walk at seven o'clock was abnormal. If he were home at seven his mother would be in the kitchen preparing dinner, sipping her vodka and tonic while his father read the newspaper and fussed about Harold Wilson. At seven o'clock, when Jenny had come and found him in his room to give him the paints, looking so sad, his parents would think: far, far too late to set out.

What were the German and the American doing? Were they so thick-skinned that they couldn't sense how much Bill hated them? Adam watched Jack and Susan slouch up to the door, giggling, the last to appear for dinner. Adam rubbed his hand up and down the thin bark. He wondered what he was missing. He didn't want to be left out. He'd backed out of the door the second he laid eyes on that awful pale blue shirt, to give the German the message that no one wanted him around.

'Ah ha! Left the bones and took the jewellery,' Susan pronounced with her mouth full of Spanish omelette. A good spanking and bed without supper is what she deserves, Laura thought, fed up. Susan was relentless – even now when they were all so tired and upset.

'There's plenty more sieving. Who knows what I won't find?' Laura retorted.

'How can you carry on with your ankle?'

'I have every intention of carrying on.'

The frayed silence that ensued continued when Adam finally pushed open the door. He slumped into the empty place. His portion of omelette was passed down from the top end of the table where Jenny was serving up seconds. The silence congealed. After the omelette Margaret had made a leek and tomato salad with a yoghurt dressing, and after that apple pie. They were like a condemned gathering grabbing its last meal.

Margaret couldn't sit down since to keep up she had to produce the apple pie the minute the rest had helped themselves to salad, to satisfy the first served who were stacking their plates. Jack and Susan and Mary Elizabeth were always the first to finish. But Margaret couldn't bear them to miss the pie.

Bill was the only one who was still picking at his omelette when Jack had finished his apple pie. When Jack stood up to go, Bill told him to wait.

Both Bill and Laura were worried that Christopher still had on the green shirt he'd worn now for three days. It filled Bill with remorse. Laura wondered if she dared offer to wash it.

CHAPTER 23

The weather really had changed. Laura shivered in the cold, starting out long before the others to hobble along the dusty road in the dark morning. Michaelis had made a stick for her out of a fig branch. She found it by her door when she got up. It must have been him, the smooth, straight grey piece of branch a perfect thinness, a curved bit at the thicker end for the handle. It did well. She was grateful, gripping it now for support. The gloomy weather was too bad, on top of everything else, presaging the winter ahead back in London where it had probably already been cold and wet and miserable for weeks. It consoled her to think of John hurrying through the rain in wet shoes.

Christopher had still not changed his shirt. Laura noticed the green collar half covered by a brown guernsey which she'd never seen before. When he caught up with her he held himself back in the awkward obvious way of a child; she watched the heavy brogues bump into each other, Christopher, with his hands at his back, frowning at the grey road.

'Don't wait for me!'

Christopher nodded, cleared his throat and rushed on. Laura smiled. It didn't matter. But when she found Michaelis at the top of the hill with an empty wheelbarrow, she was shocked.

'You're supposed to be with the director!' she barked, squeezing the smooth handle of the stick. Michaelis left the wheelbarrow and helped her to the flat stone under the olive tree where she could rest before she started sieving – which embarrassed her.

Laura's predicament with Michaelis was made worse by the fact that she disliked giving offence. She couldn't bear knowing someone hated her. Could anyone? Some people pretended it didn't matter. But if they accepted that they could be hateful, didn't it justify them in hating back? If they'd apologised or smiled or sent flowers they'd have been forgiven. Should she go over and thank Michaelis for the walking stick? Did she need to? It was only a guess that he'd made the stick for her. But who else would have? Certainly not Christopher who was too down-at-heart even to change his shirt – her sprained ankle was the least of his worries. Bill? The very thought made her laugh.

The hard stone under Laura began to gnaw into her hip bones. She stood up and looked round in the morning gloom for Christopher. Bill had said last night to expect the Epimelete towards the middle of the morning. Did Christopher want her to carry on sieving, or should she be sweeping the bottom of the robbed tomb so that the Epimelete could see better how well it had been emptied?

If she'd offended Michaelis, she'd have to name a day to go up into the mountains. Next Saturday?

Laura limped across the rough ground with her collar up against the cold. The clouds overhead blotted out all distance and sealed in the cold good and proper.

Edward and Michaelis were at the first trench she reached, but no Christopher. Nor did they know where he was. Laura swished her stick like a sword when Michaelis stopped picking. Michaelis shouted, '*Siyyia!*' Laura called '*Siyyia*' back and returned his smile.

'It's terribly swollen.' Edward frowned at Laura's ankle.

Laura waved her stick and left, reassured on Michaelis's account.

But where the hell was Christopher? Her ankle was hurting. Laura hopped over the stones, which was exhausting. Already she needed to sit down again; her arm and her back ached and the palm of her hand was red and sore. Soon there'd be a blister. What would she do then? By the time Laura reached the other trench, which was down the hill on the other side, she was weak and out of breath. Christopher was there with Susan and Manolis. He hardly looked up when Laura called. An old terrace wall was her last obstacle which she slid down on her bottom, a cascade of pebbles skittering into the trench.

'Stop that!' Christopher shouted.

Laura turned away. She stood quite still so that they wouldn't guess she'd burst into tears.

'If your ankle hurts that much, you should have stayed down at the pot shed,' was Christopher's follow-up remark, addressed to the back of Laura's blue anorak. How did he know? Laura faced him with her bleared face.

'I'm quite all right, thank you.' Christopher's expressionless stare was that of a boy who was desperate to vanish. Laura pulled a handkerchief out of her anorak pocket and blew her nose, pitying him. 'Just tell me, am I to sweep or sieve?'

Laura could sense, also, that her presence was an interruption. It mortified her to acknowledge how dispensable she was. She was angry at herself for instantly pitying the man when he'd been so rude. She must insist on her worth. Laura blew her nose until her ears popped to give vent to her hurt feelings.

'We've found another tomb.'

Laura opened her eyes.

'Probably unrobbed.' Christopher was beside her, handing her his flask of brandy. 'That's why you couldn't find me.'

A strange hush fell on all of them, Susan with her hands on her hips kicking idly at a rock while Christopher pointed out to Laura the same chunks of white marl in a tumble as they'd found before, covering the trench two feet down from the surface. Manolis stood by, holding his pick like a hat as he listened.

'The Epimelete won't leave us alone now. You can be sure of that,' Susan jeered. 'He'll want to take over, you know.'

Costas Pavlidis knew why the Ephor had sent him to the English excavation. But whatever Sideris Koraes did, he'd be arrested when news of the robbery reached the Ministry in Athens. The Colonels had been wanting to get rid of Koraes for a long time – ever since he had refused to give permission for *son et lumière* and a thousand-car car park at Phaestos. In front of a crowd of people Koraes had made a fool of Pattakos, shouting at the Colonel that as Ephor it was for him to give the permission and he refused it in the interests of Greece's posterity. It was stupid, but a man that arrogant and vain deserved to fall into such pits. A typical Chiot, over-Anglicised and too fond of foreigners! Costas on the other hand was a mainlander fron the tip of the Peloponnese and was suspicious of all islanders. He knew how proud and independent most of them were. Costas Pavlidis was a true patriot, son of the village mayor. He was grateful to the Colonels for rescuing Greece from the corrupt hands of sycophants and traitors. He had been finishing archaeology at the university when the Colonels came to power. His loyalty was intense, the young puritan that he was disgusted by the disorderliness and ineptitude of the governments before the coup. He entered the Archaeological Service full of zeal to wrest from foreign institutions as many of the ancient sites as possible. Crete was his first posting, which he didn't like.

Pavlidis was fat, with straight brown hair which would not stay brushed back. He wore dark glasses to counteract the childish character of his fine hair, also hoping they gave him the sex appeal which his sallow face and paunch lacked.

When the Epimelete's small white Fiat lurched out of the dusty track on to the new, smooth stretch of concrete, he geared down to second and inched forward, despising the cracked walls and cluttered doorways of the hovels he passed. A village boy himself, he knew and hated the filth and the meagre rooms inside with their bare floors and peeling walls, snippets of patterned oilcloth the tawdry embellishment of such poor homes. There was no dignity in this old way of living; it was unclean and unhealthy. He braked, just missing a chicken which scampered clucking in front of his wheels. The old widow whom he asked for the police station screeched at him to turn right at the crossroads. He thanked her and put the car in first.

'Who are you?' the balding officer shouted at Costas from across the table where he was sipping a coffee. The Cretan accent grated on Costas's ears. Costas, in the educated Greek he was proud of, explained his position at the museum and his business with the English archaeologists.

'Passports you want?' the officer shouted back, scattering the small pile of documents at his elbow with open-handed bravado. Costas looked away. Would someone, he asked, be so kind as to show him to the cemetery site? He did not smile; nor did the officer.

The Epimelete's arrival on the hill made everyone nervous. Costas was surprised that he could have such an effect on the English. Christopher Bendick, who Koraes had told him was in charge, fidgeted the whole time they talked, his eyes either fixed on the ground or searching the clouds like an ascending Christ. He looked tired and sad. The fact that Costas's presence made him uneasy was interesting. With rising suspicions and confidence, Costas asked to be shown around. He led the way, following his eye.

'What's this?' Costas stood over the first tomb. Christopher jumped down into the chamber and showed where they'd found the gold.

'Where is it?' Costas surveyed the lie of the land. His eyes moved restlessly behind the dark glasses while he tried to plan his next move.

'With the police.'

'Okay.'

The two men fell silent; Christopher followed the Epimelete with head down like a boy in trouble.

'What are the women doing?'

'Sieving.'

'Found anything?'

'A faience bead.'

'Nothing else?'

'Not yet.'

'The robbers did a good job!' Costas squinted at the tip of fine earth in front of Laura and Susan. He passed his fingers through it. 'You are sure nothing?' he asked the shorter woman, who'd set the bucket down when he came up. The other, seated woman was beautiful, he noticed. Her leg was in a bandage, propped up on a stone. Shy of her attractiveness, Costas looked only at Susan. 'What do you think you'll find?' he asked her. 'Aren't you too late?'

Susan glanced at Christopher. Costas followed her glance. This Christopher Bendick was such a typical Englishman, his reticence and unsmiling eyes a wall. Costas would force the man to speak. He needed to be either friends or enemies with him. The frosty no man's land which Englishmen preferred was not his cup of tea.

'You're tired,' Costas told him. 'Your face.' Costas held his hand up to his face. 'I can tell.' He took off his glasses and wiped his eyes with his handkerchief. Using the handkerchief to clean his lenses, he asked

Christopher whom he suspected. To ask him in front of the women might force him to answer. 'You suspect at any rate? Or have you no idea? We must find the evidence to prove.' He caught the eye of the beautiful, seated woman. He laughed and pointed at the heap of unsieved earth. 'You like sieving?'

'The schoolmaster thinks a German tourist had something to do with it,' Christopher broke in. 'He told that to one of our workmen.'

'Only because he hates Germans,' Laura contradicted, amazed at Christopher. Costas pushed his dark glasses back onto his nose and put his handkerchief away. He turned back to Christopher. 'Which workman did he tell this to?'

'On the matter of Germans the schoolmaster is quite insane,' Laura rushed on. 'Unfortunately this particular German one evening affronted all of us with a small frog which he threw on to our table. The schoolmaster was there. Do you remember, Christopher?' Costas recognised now the typical hysterical irrelevance women always threw in the way of discussions. It was to draw attention to themselves. They seldom understood what was being said, which was why they would interrupt.

'You've not told me yet if you like to sieve?' he asked her as he moved away with his shoulders hunched. 'You must find it very boring,' he answered for her, now stepping ahead of Christopher.

Christopher introduced Costas to Michaelis. He asked the redhead to tell the Epimelete what the schoolmaster told him. But Costas could see that the redhead was afraid. He would not look up. 'Where are you from?' Costas asked him.

'Ierapetra.'

'Have you worked on many excavations?'

'My first.'

Costas would deal with the workmen at another time.

Costas now wanted to learn from Christopher about the workman on guard when the tomb was robbed. He was sure there was someone, although Koraes had claimed he knew nothing since Bendick had not mentioned anyone. It was strange Koraes had not asked Bendick. He suspected the Ephor was so frightened that he couldn't think clearly, which put the responsibility for discovering the robbery completely on to Costas. There was a cold wind on the hill. Costas shivered and rubbed his hands together. Bendick offered Costas brandy from a silver flask.

Flattered and grateful, Costas thought that after all he could like this aristocratic Englishman. He drank to Christopher's health, accepting two more tots of the warming drink; with the third he wondered if Bendick might not become his friend.

'I've read your book,' he told him.

'Oh. Good.'

'I like it. It's good.'

Christopher put away the flask and offered to show the Epimelete the pot shed, where it would also be warmer.

'You're cold?' Costas asked.

'A little.'

'Let's go then.'

It was on the way down the hill to the road, Christopher now leading the way, that Costas asked why the night watchman had not caught the robbers, if there had been a night watchman. Had there been a night watchman? Because surely the night watchman would know who the robbers were. But Bendick did not hear.

Susan dumped another bucket of earth into the sieve. Laura snapped at her not to be in such a hurry.

'It's hopeless. Why don't you quit?'

'Those beads weren't bad. Christopher was pleased.' Laura caught Susan sneering down at the heaped-up sieve. Stung, she looked down again at what she was doing, slowly pushing her fingers through the earth until she felt the rough mesh at the bottom. It rankled. Quite unjustified and unsportsmanlike. Susan was an unforgiving, wintry soul. Laura wished Christopher hadn't ordered her to help. He'd made a quick decision down at the new tomb, which had surprised all of them. Abruptly he'd sent Manolis to help Michaelis and Edward, and told Susan to put away her notebook and get on with sieving. When Susan argued he told her to shut up.

'Christopher's a fool,' Susan retorted. 'He's in for it.'

With growing uneasiness Laura pushed her hand faster back and forth. Susan was an archaeologist, and therefore would know what she was talking about. Her threats confused Laura, who wanted to stick up for Christopher but who was afraid to speak out because she didn't know enough.

'Let's get on then,' Laura urged, hunched over the sieve like a refugee.

Susan shovelled more earth from the other pile into her bucket and lugged it across. 'Ready?'

'Fire away,' Laura sang out.

Susan threw down the empty bucket and began to make patterns in the dust with the olive root Laura had just discarded from the fresh load.

'You could help me,' Laura suggested.

'Could I?'

'It would go faster.'

With a groan Susan got to her feet. Laura was right; her hands pushing through the earth along with Laura's made it go a lot faster.

'What would you like to find? I mean, if we're lucky and find something,' Laura asked her.

'Think what the robbers got! It's sickening. We find three faience beads. The robbers found whole necklaces probably. And juglets of faience. What else? A gold cup. We'll never never know what was in this tomb now. Christopher's hopeless. I must find out at what time of the day he was born. He should never have left Andonis here the other night.'

'But if it was in his stars, he couldn't help it. Or can you help what's in your stars? Look! Susan!' Laura held up in the palm of her hand a long gold pin with a small knob at the end. Carefully Susan took it from her.

'I'll show the others.' Susan disappeared over the brow of the hill at a run. Laura heard a few minutes later a low faraway exclamation dulled by the cold wind and grey sky.

Laura was pleased. She'd cheered them up.

But down at the pot shed Jack was not cheerful; he was shooting rubber bands at the frosted glass shade of the Camping Gaz. 'Ping' it went every time Jack got a bull's eye, which pushed him on to shoot another and another. When the box of rubber bands was empty Jack climbed down from his stool and crawled under the table to recover his ammunition. The soles of his tennis shoes and the bottoms of his faded jeans were all that Christopher could see when he pushed open the door.

The Epimelete followed Christopher into the room.

Christopher called. Jack scrambled up from under the table, still with his pipe clenched between his teeth. He dusted himself off and took his pipe out of his mouth to offer his hand to the Epimelete. If he hadn't felt caught out he wouldn't have been so polite; he was not usually one who shook hands with strangers. But for once he was courteous and smiled, ashamed of himself.

'Our draughtsman,' explained Christopher. When he lifted from the pile of notebooks in front of Jack's drawing board the model of the bird, to hand it to the Epimelete, his eye happened upon the drawing tacked to Jack's board. It was not of the Middle Minoan bird. Christopher leaned closer. He squinted at the exuberant drawing of a bull game on a wide beaker-shaped cup.

'We must catch the robbers, if this is the sort of thing.' The Epimelete had set the bird back on the notebooks and was looking around the room.

'Recognise it?' Jack asked Christopher when Christopher would not unglue himself from the drawing of the beaker.

The Epimelete thumped the board. 'Congratulations! I must see it at once.'

'I made it up.'

The Epimelete turned on Jack. 'You did what?'

Jack shrugged. 'I invented it. The robbers could have found something like it.' Jack sucked at his pipe. 'You like it?' he asked the Greek official, feigning a playful detachment.

More angry than Jack realised, Christopher untacked it and scrumpled it up, tossing the ball of paper into an empty box under the table. Jack pulled out his drawing of the bird and tacked it on to the board. While the Epimelete grinned at Jack, Christopher checked Jack's drawing with callipers, measuring the beak, the head and the wings of the model against Jack's pencilled outline. Jack waited for the verdict, but the Epimelete, who was more his age than Christopher's, began to jingle his car keys. Jack asked if he wanted to see more of the finds; the Epimelete asked Jack what had given him the idea of the bull game which was 'amusing'.

'Perhaps our Egyptian import would interest you,' Jack persevered. He lifted down the cardboard box and pulled the faience bowl out of cotton wool. Christopher looked up. The young round face of the Epimelete lit up with exactly the pleasure Christopher had felt when Susan showed it to him just after she'd found all the pieces. Christopher put down the callipers and went over to look at the bowl again with the Epimelete, screwing up his eyes as if the glory of such a find could hurt.

'Look there! The fishes in the pond.' The Epimelete was bending so close that his nose touched the rim. 'That's the lotus. You see?'

Christopher nodded. 'Eighteenth Dynasty, I thought.'

'Definitely. You know the ones from Deir el Bahri?'

Jack rummaged amid the pile of tapes and string on Adam's table for the raki bottle. He asked Costas if he'd like some. Christopher declined but the Epimelete looked willing, so Jack absented himself to wash a glass outside under the hose.

Ellen cowered in the yard before a table of sherds, wearing a hideously garish red pullover. Until then Jack hadn't noticed the change in the weather. He scanned the lowering clouds for the first time and suggested she knock off for a bit.

The Epimelete had taken off his dark glasses. He and Christopher were leaning on Adam's drawing board, discussing faience. Christopher looked awkward and entranced; the Epimelete shook a hand as he expatiated on faience from Sinai. Such intensity would have put Jack off. But Christopher's new-looking guernsey looked even newer as he sat so still listening.

The creak and judder of the door which announced Jack's return with the clean glass also let in Ellen. When she clapped eyes on the Epimelete she asked who he thought had robbed the tomb.

'It would be a help to know, although knowing wouldn't necessarily retrieve the objects, since they may already have left the country.' Ellen stood waif-like in her bulky, bright sweater.

Both men remembered themselves. Costas put on his dark glasses and straightened the bottom edge of his leather jacket. Christopher dropped his arms. Then he was pecking at the bird model with the callipers again, and although he listened he pretended not to. Did Christopher realise how predictable he was, Jack wondered. Jack handed Costas the glass of raki and sat down on the rush-bottomed chair under the window.

What angered Adam was not Bill's jitters but his reticence. All morning up on the Kallithea site Bill darted about from trench to trench in a silence. He had his notebook and was jotting things down, but absolutely no one did he take into his confidence. Not even the beautiful Mary Elizabeth, because Adam kept watch. Which Adam thought insulting to all of them. It wasn't as if all of them weren't wondering who did it. Was Bill trying to pretend it only concerned him? All he said to Adam when he glanced at his drawing of the wall was, 'What did you forget today? Hmmm?' But Adam would have forgiven him even that if he'd admitted how worried he was.

There had been moments when Adam liked Bill, which was why his eye now followed Bill around the site, and his scrutiny of the big burly man smoking cigarette after cigarette absorbed most of his attention – his drawing of the fourth course of stones progressed slowly. He'd always liked Bill's size and his liveliness. Also his authority. He was the sort of man you assumed knew what needed doing, and knew how to make people do it. Not repressed and cowardly like Christopher. A brave man, Adam used to think, until the evening Bill returned from Ierapetra with Mary Elizabeth and pretended that the policeman was a small matter. That was vain and unkind to Jenny. When Jenny had called down the table asking if the policeman carried a gun, Bill ignored her. That came back to Adam now – Bill carried on talking to Annabel as if he hadn't heard. It was Mary Elizabeth who had forced him to tell the story. Could anyone force Bill now to climb off his perch? Bill was in a black anorak which Adam had never seen before, with a large zip-up collar pulled up around his neck. The black anorak against his black hair darkened him dramatically, appearing to remove Bill still further from the green and blue and yellow sweaters the rest of them wore. The death was the weather, not the robbery, but Bill's attire was, Adam thought, typical. Poor Jenny. He would protect her. In his bright blue sweater (which jarred slightly with his purple jeans but never mind) he'd take Jenny for a walk after lunch to share with her his thoughts on who had planned the robbery and framed Andonis.

The tomato soup was so good at lunch that Adam asked for a third helping. Margaret, who sat next to him, carried his bowl down to the other end of the table where Jenny was ladling out. She'd insisted there was enough when Adam congratulated her. It was the first time Adam had praised her cooking; Margaret was thrilled. After the soup there was pork stew with apples and celery which was also delicious, and Adam was tempted to have a second helping of that as well. Margaret told him she'd added chervil and allspice, which provided the special flavour. The cold weather gave her the idea of the allspice, she went on, expanding on her cooking and seasoning when she saw how hungry he was.

Jenny was also wearing something he'd never seen before, a white polo-neck which made her grey hair look younger. She refused to look Adam's way, although he tried several times to catch her attention. She was wedged between Laura and Edward and was talking to Annabel about the price of Shetlands in London. Laura interrupted the conversation with the remark that Shetlands gave women fat bottoms. Adam could see Laura had wanted to annoy Annabel and had managed it. She was in high spirits because of the gold pin. She'd pressed it on them the minute they came through the door. Jenny burst out laughing, looking suddenly soft and ravishing in the white polo-neck; Adam felt tears spring to his eyes at such joy melting his Jenny's face. Had Bill noticed? Adam checked. No, he hadn't. His eyes were on his plate of stew.

'Leave room for my sherry trifle. Or I'll be most offended,' Margaret was saying. Adam nodded. His hunger had left him.

The noise of shuffling and shouting in the hollow room made it hard to hear. The telephone was at the back of the Merry Widow café, by the low door into the kitchen, set on crates beside a basket of onions. Christopher blinked back tears from the onions, crouched down because the crates were so low. He covered his other ear to try and hear. Unfortunately Andonis's wife was not used to the telephone and didn't realise that screeching into it made gibberish of her answers, forcing Christopher to shout even louder.

'Have you understood me?' By now he was desperate. 'That he must catch a bus *today*? Today, Kyria Maria.' At last Christopher could make out a 'yes, yes, yes'. He put down the telephone and painfully raised himself off his haunches, rubbing his aching knees. Thank God he'd got through to Maria. She was a dependable woman who would get Andonis on that bus.

Christopher was astounded when the Epimelete asked on the way down the hill if there had been a night watchman. Why hadn't Koraes given the Epimelete Andonis's name? Koraes had shown such distrust over the telephone when Christopher reported the gold finds. Now the robbery proved that distrust. Why hadn't he had Andonis arrested? And why hadn't he turned up himself to organise better guards? Christopher had dreaded the arrival of the Epimelete. He stopped digging the new trench: he sent Manolis to help Edward and Michaelis, and Susan to sieve with Laura, because he expected the Epimelete to take over the cemetery site – which he was damned if he'd let happen. It surprised him that he would hide from the Epimelete the next unrobbed tomb. But he had to show his mother some time.

Now, though, discovering that the Ephor had not even mentioned Andonis to the Epimelete confounded him. And when he had shown the Epimelete the new trench, the Epimelete asked no questions. To add to that he turned out to be as thrilled as Christopher by the faience bowl. Costas Pavlidis wasn't at all a bad sort. The only danger was his harping on a night watchman's evidence. He was right, which was why Andonis must come back to defend himself. Christopher was sure Andonis had abandoned his post by the time the robbers arrived. He would have nothing to tell. But Andonis was safer admitting this to the Epimelete in Christopher's presence than at a police station in Heraklion. Christopher was grateful to Koraes although he didn't understand him. The foreman invited Christopher to join them for a raki as he walked out. He gave Christopher a conspiratorial look. Georgos pulled up a chair, but Christopher declined.

At headquarters everyone was on the trifle by the time Christopher squeezed through the door. Laura called down the room that she had something to show.

CHAPTER 24

As the English archaeologists walked past his 'government' café, the mayor had pointed them out to the policeman who kept out of sight inside, peering through the café window. The mayor offered to help Sotiris Georgakis in any way he could but he demurred when the policeman expressed his opinion that the archaeologists were 'a bad lot', for, although the Colonels' man, the mayor had a good heart, and was proud of what the archaeologists did for the village. He shrugged when Georgakis could report from studying the passports that the director

Bill's wife Jenny was nineteen years older than the pale chap Adam who kissed her in the street; and that the beautiful American girl who had been with the director was born on 12 May 1948 in Charlotte, North Carolina; the director Bill twenty years older than she, born in London on the 16th of August, 1928. But the mayor thought Georgakis was wise to expect to hear lies from the English about what they were doing the night of the robbery.

The officer had given Sotiris the job of finding out where each one of the English were on the night. Sotiris put off by a whole day his visit to their headquarters. First he'd found out at what hour they collected for meals. And their dates and faces he fixed in his mind, with the help of the mayor, who knew Georgakis was doing an important job. Sotiris had even inspected the different houses where the archaeologists slept when they were not there – thrilled by the beautiful orange nightgown strewn across one of the beds where the women Laura and Annabel lived. It was a nightgown of sex, and when the mayor pointed out Laura, who limped past with her bandaged ankle, Sotiris gave a triumphant nod at the window.

Sotiris knocked on the door and waited. The hubbub inside drowned his knock. He knocked again. The English lord type called Christopher opened the door. Sotiris kept his cap on, demanding to be let in. Christopher opened the door wider to let him through. Abashed by so many foreign faces turned on him, Sotiris pulled down his cap and fumbled in his pocket for his pad and pencil. Christopher offered him a chair, which he refused. The noise of talk had gone. They waited for him to speak. Then the grey-haired Jenny appeared in the doorway at the far end and let out a short scream before clapping her hand over her mouth. This encouraged Sotiris. He picked out the Adam chap among the faces at the table, and was gratified to see him after that fix his eyes on the director's old wife. No one else looked at her.

The policeman held up his pad and pencil. 'We must know what each one of you was doing on the night of the robbery. You all must stay until I write it down. This is a police investigation.' They hung their heads like naughty children. Sotiris pulled their passports out of his other, bulging pocket and slapped them down on the table. The first one was green. 'Mary Elizabeth . . .' He couldn't pronounce the last name.

'Johnson,' put in the director Bill.

'Where were you on the night of October the 22nd which was the night of the robbery of the tomb? Tell me!' he ordered her. He blinked. 'And it must be the truth because this is a police investigation.'

'Asleep.'

Sotiris thought he heard a giggle and ran his eye up and down the table. But still their heads hung down, and no one seemed to be moving their lips.

'Were you asleep with anyone?'

This time he heard a laugh, and when he looked up saw it was the short girl with curly hair called Susan. 'She was asleep with me and her and her,' Susan told him, still laughing. Susan was nudged by the blonde girl next to her; she hung her head again and pulled in her lips. He gave Susan a severe frown which he thought should teach her a lesson.

Suddenly behind him the door creaked and scraped and nearly knocked him over. The hairy Englishman called Jack and the Epimelete from the Heraklion museum pushed their way in. Sotiris dropped his pad and nearly lost his cap when he leaned over. Chairs and food were produced for both men who sat down with the others. The policeman waited, patting his pad impatiently.

'What is all this anyway?' The Epimelete interrupted, just as Sotiris was about to continue.

'A police investigation.'

'Pshaw. What about?'

'The robbery of the tomb on the night of October 22nd.'

'But that's no police matter. That's why I'm here. I'm from the museum.'

'I know,' retorted the policeman.

'How do you know? I've never seen you before.'

Sotiris Georgakis pulled up his chin, feeling the collar of his jacket rub against it. 'I know who you are. I know who everyone at this table is.' Sotiris wagged his finger. 'He is the director Bill born in London the 16th of August 1928! She is his wife Jenny born the 21st of January 1929. In London also.'

'And when was I born?' asked the Epimelete, looking round the table for smiles. He slipped his ID card out of his wallet and thrust it at Sotiris.

'You aren't an archaeologist,' the policeman defended himself.

'I *am* an archaeologist.'

'You are a government official. Like me.'

The Epimelete laughed. He waved the card at Sotiris before he put it away and forked into his mouth a huge bite of stew. 'You Cretans,' he continued with a full mouth, 'understand nothing. You're all such heroes!'

Sotiris blushed; veins in his neck started up against his tight collar. 'I am doing my job,' he shouted.

Adam hit the table. 'If you're doing your job, do it somewhere else.'

'That could go for all of us,' Jack mumbled, which Adam heard. Adam jumped to his feet. 'This is the little Hitler who aimed his gun at Bill.' Adam nodded at the Epimelete. 'A hero. Some *palikari*! He frightens people. He's a fool, a *malakas*. Get him out of here!' Adam screamed and sat down; he'd gone white. Jenny mimed to Margaret from across the room to give Adam a drink.

The policeman grabbed the pile of passports with both hands and walked out.

The clink and creak and dull thud followed, as always when the dust settles after a crash. Susan and Ellen were snivelling into their paper napkins. Margaret's wooden sandals clacketty-clacked into the kitchen with Adam's empty glass. The Epimelete made a lapping sound as he chewed with his mouth open, and slurped down his wine.

Jack and the Epimelete were the only ones still eating and had more meal to look forward to. Jack had seen Christopher's sherry trifle and was determined to have some.

'Well,' Jack prattled, 'he tried. Poor chap really. He was only doing his job.'

'It's not his job,' argued the Epimelete between bites. 'Only eventually when we know who did it is it a police matter as well. He was . . . how do you say . . . premature.'

'Precipitate.'

'Precipitate, yes. He was!'

Laura was looking askance at the Epimelete's eating habits, and decided he should be told the English attitude to policemen who carry guns. Her embarking on this 'deeply held view' gave Christopher and Bill the chance to get up and leave. Jenny sat on, listening intently, soothed by Laura's anger and eloquence.

'I know,' interrupted the Epimelete. 'I've been to London, Oxford and Nottingham.'

'Well,' Laura persisted. 'You can understand then.'

'I was terrified,' Jenny admitted.

'He pointed it at you?' the Epimelete asked.

'No, but he looked as if he might.'

Both Jack and the Epimelete laughed. Jenny ducked her head. Adam slouched into his chair at the other end of the table. Half-closing his eyes, he pictured Jack bent backwards over a barrel with his tongue ripped out, as he picked at the corner of his paper napkin.

Clacketty-clack, Margaret came back with two helpings of trifle. She took Jack's and Costas's empty plates away, reminding Susan that it was her turn to wash up. Laura lifted her fig stick off the back of her chair. Costas congratulated her on choosing fig. Bowing from her waist and flourishing the stick, Laura thanked the Epimelete for the compliment. The Epimelete's chuckle did not cover up his embarrassment. Laura laid a hand on Adam's shoulder as she passed his thin, slumped back. He was brave, and very English, she was thinking, in the way most people underestimated him and did not realise the passion for justice burning in the boy, a hot blue flame of real feeling.

Jack and Costas Pavlidis were both delighted that they had struck up a friendship. They were late to lunch because they'd been talking at the café. Costas was surprised by his fluent English, which he had little chance to use in the museum; only when colleagues visited and the Ephor was too busy could he look after them and converse. He could also understand, although Jack spoke quickly as if Costas were an Englishman. Costas was pleased.

Christopher had come with them, ordering the first round of beers from Vasilis. But then Christopher had had to go to the telephone, and Jack and the Epimelete had sat on, with Vasilis hovering. Jack and Costas relished Vasilis's frustrated curiosity since they spoke only English.

When Christopher left, Costas asked Jack if there had been a night watchman. Jack smirked and ordered two more beers. 'But hasn't Christopher told you?'

'Why does that amuse you?'

'He doesn't want you to know that Andonis Markakis disappeared. He was the night watchman and by the morning he'd gone.'

Costas ripped off his black glasses. 'Why has no one told me?' Jack shook his head. 'Christopher's sticking up for the underdog. Andonis is his friend. He must stand up for him. That's why.' Jack leaned nearer. 'Actually, it was the cook Margaret who told me a very funny thing. Yesterday evening after the robbery, when everyone collected for dinner and a German tourist with his American sidekick were grilling Bill on what he knew because the schoolmaster was blaming the robbery on the German, Christopher came in drunk or like he had toothache. He said nothing until the German mentioned Andonis's name and asked if the old man had done it. Suddenly Christopher shouted, "No, he's my friend!" Bill didn't know where to look. His co-director, behaving like a crazy man . . .'

'Whose friend?' Vasilis shouted from the next table.

Costas ignored Vasilis, leaning forward on both elbows. 'You amaze me,' he murmured.

'What does "amaze" mean?' asked Vasilis.

Suddenly the Epimelete hissed at Jack and Vasilis both to be quiet. Jack was confused – to be so rudely interrupted by his new friend! 'The old workman did it,' whispered the excited Epimelete. 'Dowries, the new television, the refrigerator he can't afford. We prove Markakis did it and the matter's settled. Your Christopher is insane.'

Vasilis blew his nose onto the ground and wiped it with his hand; then he cleared his throat and spat. When both Costas and Jack looked up, he twirled round in his chair and walked away. Costas shook his head, sunk in his own thoughts. Why did Christopher wish to involve himself in an old Cretan's crime? Especially when it was so clear that the old man had done it. This didn't tally with his view of Englishmen. Costas had visited England. Englishmen were astute and cunning and greedy. With Englishmen you had to keep on your toes. Greeks who believed Englishmen were honourable were deceived by English mannerisms like the pipe and the understatement. Costas knew that English style had nothing to do with English morality, which was a myth. The poor Ephor Koraes he despised for just that reason, because he could not see through all those so-called English traditions.

But then here he was, hearing about a real Englishman who was naive, idiotically loyal and inept. The Epimelete was irritated and disturbed, growing angrier by the second. He started tapping his glasses on the wooden table.

Jack drank down the rest of his beer in two gulps, miffed that he'd been told to be quiet by a ruddy local. It was a bit off. Jack began to wonder if he'd said too much. When Georgos lumbered through the door in his patched grey trousers and Norwegian sweater with the red reindeers, Jack called out, relieved to see him.

Georgos shouted to Vasilis to bring him a carafe of raki. He dragged over a chair, scraping and bumping its legs until he could collapse into it and fold his hands in his lap. Maria brought the raki and a quartered pomegranate. Jack asked for another beer. The Epimelete raised his head at last but declined Jack's offer of another beer.

The presence of the Epimelete made Georgos circumspect; his attention swept the room before and after his first raki. With his second he asked the Epimelete where he was from. 'I'm a Cretan! I'm a hero!' Georgos slapped his chest and then roared with laughter.

'The Peloponnese.'

'A Mycenaean!' Georgos retorted. Costas finally smiled and put on his black glasses.

'Are the Mycenaeans also heroes?' he asked Georgos.

'Mortal heroes. We're immortal heroes. Sons of Zeus!' Georgos wished the Epimelete health and drank down his third raki.

The café grew noisier. Two sets of barefoot hippies came in, in frayed jeans and skimpy T-shirts. They were American; their guttural *ers* and slow slang changed the atmosphere. Both Georgos and Costas eyed them. The farmers at the next table in Wellingtons and plaid shirts left. The tractor outside coughed and chortled, making the windows rattle.

When Jack returned from paying Maria and asked Costas if he'd like lunch, Georgos told them how he'd overheard Christopher on the telephone to Markakis's wife just now, telling her to send Andonis back on the bus.

'That policeman should be shot.'

'Tomorrow. Okay, Laura? You will come?'

'I'm not talking about the mountains. I'm talking about that awful policeman.' Laura picked up a stone and threw it, hitting the trunk of an orange tree several yards away. The trunks looked blacker in the dusk, and closer, the chattering of birds overhead in the dark blue sky frantic; night was near. 'I must get back. Someone will see us.'

Michaelis looked so provoked as he glanced over his shoulder that Laura laughed. Michaelis raised her hands and kissed them. They were alone. It was almost night. Everyone had gone home.

Laura sighed.

'What's wrong?'

Laura pulled her hands away and retrieved her stick from where it had fallen at their feet. Slowly she started to hobble back toward the road. She wanted to get home, but didn't quite know how to manage it gently. The danger of offending Michaelis intensified by the minute. He'd waylaid her as she was scrabbling down the rubbly hill on her way to the loo; he'd be waiting for her in the orange groves, he said. She'd had to come. And now it was the mountains, mountains, mountains. What on earth could happen in the mountains that couldn't happen anywhere else? Was Michaelis some kind of romantic? Laura had always thought Mediterraneans practical people who didn't indulge in such fancies.

'I haven't thanked you for this stick,' she called back. 'It's fantastic.'

'It's a fig.'

'The Epimelete congratulated me on picking out fig for a walking stick.' Laura grinned. 'I didn't tell him it was you.'

'The Epimelete wanted me to speak, but I won't. He's a spy.'

'What?'

'Never mind.'

Laura had reached the road and Michaelis was just behind.

'Anyway, thank you,' Laura said.

'For what?'

'This stick, you fool.'

'But I didn't give it to you. Why thank me?'

Now it was Laura who looked provoked. She stared down at the thin white thing which had already become such a friend. 'But . . . I thought . . . leaning there waiting for me . . . Who left it then?' she stuttered.

Michaelis shrugged. 'Christopher perhaps. He likes you very much.'

Laura blushed, which it was too dark now for Michaelis to notice.

'No, it wasn't Christopher,' Laura retorted, digging the stick under a root which the road had exposed. Poor denuded root, Laura thought inconsequentially. With their backs to the village they followed the white line of the road, walking towards a black blankness of mountain. Sprinklings of lights in the void signified a village with electricity.

'I shall show you the running stream of beautiful water and the walnut trees,' Michaelis was saying.

'When?'

'Tomorrow?'

'Not tomorrow. The weekend. What about Saturday?'

Michaelis laid his hand on Laura's neck and kissed her ear. 'I have taken a room in the village,' he told her.

'You've what?'

'I'm sleeping in the village now. Ierapetra's too far every night.'

They stared into each other's faces, their eyes and hair as dark as the dark hills, only the forehead and cheeks pale. 'Well then, let's go. There's still time before supper.'

'Right now?'

'If you like.' Laura laughed, switching her stick to the other hand so that she could take Michaelis's arm. They turned their backs on the mountains; the village lay ahead now. They walked towards it very slowly like a long-married couple.

'I thought Laura was marvellous. She really told the Epimelete!'

Adam shivered. He clenched his teeth and fists, rigid with anger, huddled under Jenny's rug. 'I hate that man. I hate him, I hate him, I hate him.' He took in a deep breath and let it out in a huff. He took in another and another. 'He's a right shit,' he let out, calming down.

The sea was calming as its flat, colourless cold lapped against the rocks. All around them it was grey: the cliff, the sky, the beach, the sea which was a glassier and darker grey, grey like everything else. Adam was glad Jenny had suggested walking this far. No one would see them here. She'd brought a rug from her car, anticipating how cold it would be.

'Laura's on our side.'

'I wish we could go to the land where the bong-tree grows, for a year and a day . . .'

'They sailed away . . . But you must admit, I'm right about Laura.' Jenny slid her arm through Adam's and gave him a squeeze.

'She's a sight better than Jack.'

'I agree with you.'

'To put it mildly.'

'Oh! I thought you meant the policeman just now.' Jenny slipped both arms around the boy and hugged him, the rug like a shelter from the storm. Adam's hate was like a storm, thought Jenny, the poor boy. And he smelt good. Of tea leaves and cinnamon. And mothballs. 'You smell of mothballs,' she complained.

'Mum buries everything in mothballs to preserve them.' Adam startled Jenny with a lewd chortle.

Jenny pulled away.

'What's wrong?' Adam twisted round, picking up a handful of pebbles.

'I'm cold.'

'Let's go back then.'

'No.'

'Tell me.' Adam leaned his peeling nose and small brown eyes closer. He picked up her hand and felt her knuckles with his nervous fingers, kissing her on the forehead. Jenny couldn't look.

'Has Bill talked to you about the robbery?' Adam asked, still fingering her hand.

'No.'

'I felt almost sorry for the policeman at lunch.'

'Oh?'

'The Epimelete made a fool of him.'

Jenny threw off the rug, and stood up. 'Why'd you shout at him then?'

Still seated, Adam looked up into her face in the unguarded, pitiful way of a trusting five-year-old. His wavy blond hair was lank in the grey cold; the hand she'd just discarded lay palm upward on the pebbles. 'I could see how terrified you were,' he said. 'You looked like an angel in your white sweater. I'd never seen you in white before. You looked beautiful, but when the policeman came in, he changed you into a frightened old woman. I couldn't stand it. It was for your sake I shouted at him. To bring back your youth and your beauty. I was actually feeling rather sorry for the man when I did it.'

The young! What was she doing on this beach with Adam? Jenny gripped her head and ran down to the water's edge. She tossed pebbles into the ugly water and wished that Adam would take her into his arms and kiss her as he'd kissed her yesterday, so that she could forget. Everything. And not think . . . The washing! She'd forgotten. She had to go and collect it from the fat woman. Oh God. Damn. And she'd have to gulp down another one of those sweet pastes.

Adam's arms were around her. He was nuzzling into her neck, his fingers on her breasts, cupping them in his hands. Oh, she craved it. In a swoon with half-closed eyes, she turned and they kissed.

Bill was too impatient to consider Christopher. When there was no answer he slapped the door with the palm of his hand. Since there was no one to see, the pot shed empty, neither Jack nor Ellen nor any of the others studying the pottery or catching up on their notebooks as they ought to be doing, Bill crouched over and pressed his ear to the door. A strange diffidence had crept into his relations with Christopher during the last twenty-four hours. He slapped the door again. And again he crouched over and pressed his ear to the wood. At last he heard the faint flop of a paperback dropped to the floor, and a mumbled 'Yes?'

'Where is everybody?' Bill shouted, barging into the room. Christopher lay with his feet up, his hands behind his head. He stared up at his tall friend-enemy-whatever and shook his head. 'They should be down here working. It's six o'clock.' Bill tapped the hard top of Christopher's blue suitcase. 'What are you doing?'

Christopher raised himself on one elbow and looked around the room for a chair. 'Throw the stuff on the floor and sit down if you like.' He motioned to the chair heaped with dirty pants and shirts and his khaki jacket.

'You should give your washing to Jenny,' Bill reproached him. 'She takes it to the fat woman.' The green shirt was in the heap on the chair, which irritated Bill. Christopher was not to grow slovenly under pressure. Bill depended too much on his fastidiousness.

'I'm sorry if my dirty clothes bother you,' Christopher tweaked Bill. He grinned when Bill scorned the clothes, the corners of his mouth gone right down.

'Where's the raki?'

'On Adam's table behind the tapes.'

Christopher could hear Bill muttering over Adam's mess; Christopher collapsed back onto his pillow and shut his eyes.

'I've come down here to apologise, so wake up. We can't fall out now. Let's make a plan and stick by it. Come what may we'll make a success of this excavation yet.' Bill was bent over in the doorway, clutching the raki bottle. Christopher shoved both hands back under his head. He giggled.

Bill still held the raki bottle in one hand and a pink cup in the other as he sprawled out in the chair. 'I was so angry with Andonis, that was the trouble. But it's no good.'

'I'm jolly angry,' Christopher chimed in.

Bill blinked. The polythene on the window over Christopher's head muffled the intrusive drone of a fly – in crazy, unrhythmic pauses. Suddenly Christopher bounded into the next room and rummaged about for the fly swatter. Things clunked and clattered.

'Leave it,' Bill called.

Christopher ran back in his stocking feet, brandishing the yellow weapon.

'You're as bad as Jenny,' Bill snapped.

He jumped up on to his bed; the canvas made a yonking sound. Christopher swiped at the fly and got it, unsticking one of the tapes that held the polythene to the wall.

'Pour yourself a brandy, for God's sake. And sit down!' Bill pleaded.

But Christopher, perched on the edge of the camp bed, would not let go of the yellow swatter. And Bill couldn't take his eyes off the perforated snippet of plastic and the wire handle squeezed between Christopher's knees. The loose polythene on the window now gurgled.

'I've summoned Andonis back,' Christopher announced. 'I hope you don't mind, but I thought it was best.'

Bill flinched. 'But they've arrested him, haven't they?'

'No. The Epimelete asked me if we'd had a night watchman.'

'But you gave Koraes Andonis's name. I heard you.'

Christopher shrugged. 'For some reason he hasn't told Pavlidis.'

Bill felt his hold on the conversation seep away. Christopher was beginning to frighten him. What would be the next thing? Summon back Andonis? Bill shook his head to try and clear his mind. Christopher stared evasively at the ceiling, with funny narrowed eyes. Was he drunk? Bill gulped down more raki.

'Tell Pavlidis, when he asks you, that I was the night watchman. Nothing to lose.' Christopher was still squinting at the ceiling as he spoke.

Bill gawped at his old friend. 'You mean tell him you robbed the tomb?'

'Costas Pavlidis will think Andonis did, if he hears about him. Say I came back to the village at four thirty for a quick wash and brush-up before starting work, which gave the robbers time to do their stuff before we started work at six.'

Christopher peeped across at Bill now as if scheming for seconds of pudding. He was gauging Bill's reaction. He even smiled. 'I think,' Christopher whispered, 'that we've found another unrobbed tomb, but I haven't let the Epimelete know. I sent everyone away when he arrived, so he wouldn't see.'

'Why?' Bill dug out another cigarette. His hand was shaking.

'Because I want to dig it. That's all.'

'That's *all*!' Bill took in a deep breath. He rubbed his eyes and inhaled, blowing the smoke out slowly.

'Well.' Bill spoke quietly. 'I'm not telling Pavlidis that you were the night watchman. Koraes knows you weren't and knows that Andonis was.'

Christopher frowned.

'And I think it's very wrong to withold information from Pavlidis, who is a Greek official.' Bill was still holding his bottle and the pink cup as he got to his feet. 'I must run this excavation properly.'

Christopher flapped the swatter, still perched on the edge of his bed. 'Sorry if I've upset you.'

'Of course you've upset me.'

'Sit back down.' Christopher threw the swatter on the floor.

When Bill left, there was just enough light to follow the path down to the beach. Bill needed to calm down. At least he and Christopher were friends again, but not to consult him before summoning Andonis back niggled. Christopher should have let Andonis be. They hadn't arrested him, which must mean that the Ephor knew something already which exonerated Andonis. Couldn't Christopher see that? Was Christopher afraid Andonis was the robber? Was he trying to prove a guilty man innocent because he was his friend? That was taking friendship too far!

Bill loathed the robbery. He didn't want to discuss it with anyone. Leave it to the authorities. He didn't mind at all the policeman at lunchtime doing his little investigation. He resented Christopher meddling.

It was at this point that Bill saw coming out of the dark up the path a wide, awkward shape. He must get out of the way. He stumbled, catching himself on a sharp stone; as he nursed his hand it occurred to him that if he hadn't gone to see Christopher just now, he would have thought Andonis came back because he was innocent.

The strange shape turned out to be two people huddled under a blanket, the odd shapes that had looked like horns their heads and shoulders, giggling in a low intimate way; they looked like a couple of Mexicans. Hippies, no doubt. Bill was waiting on his knees for them to pass. Off to beg food from some poor villager who couldn't help being hospitable. It was too dark to make out even the colour of the blanket, though in the daylight he'd have probably seen a gruesome yellow and blue check like the rug Jenny bought in Ierapetra, which had shocked Bill when she showed it to him, it was so synthetic and loud. But she'd thought it would do for picnics.

Bill carried on. He could hear the sea ahead. A wave crashed close by. He stopped and listened – zipping up his anorak – to the hollow, watery darkness. The sea frightened him. He was a poor swimmer. Another wave broke like a clap of thunder. Bill stiffened. It was no good. He had better turn back. And as he did, he recalled the giggle he'd heard from under the blanket. It worried him now.

When he reached the turning to the Merry Widow café and headquarters, Vasilis called, '*Kyrie* Bill!' He stood on the step in front of his café.

Bill could see little in the dark.

'The Ephor was looking for you,' Vasilis shouted. 'He's gone. He took the gold. He was here five minutes ago.'

CHAPTER 25

The Ephor went straight to the police station to collect the earrings and pins, then drove back into the village to find Vasilis Andonakis. The wife made him a coffee before she went to fetch her husband from the orange groves. The Ephor would have preferred to go with her, but she made it plain she didn't want him to, insisting she'd only be a minute. Alone in the café while the wife was away, the Ephor moved to the table behind the stove, not wishing to be recognised, especially by the English.

When Vasilis marched in wearing the same black T-shirt and smutty brown trousers, the Ephor came out from behind his cover and held out his hand. They kept their eyes down as they greeted each other without surprise. The Ephor returned to the table behind the stove and Vasilis pulled up a chair; Maria shrieked from across the room that she would bring them a drink.

Vasilis expected the Ephor to be angry, and was increasingly ill at ease when the Ephor in his smart three-piece suit remained polite and calm. Determined this time to make a better job of his visit, the Ephor played his part immaculately. The cunning café owner was nonplussed.

'I have been speaking with your cousin in Athens, the general,' Koraes began. Vasilis's mouth opened a little. 'You might have supposed I was a weak and fearful man, but you were wrong. It was my mistake to threaten you when I was here last. For that I apologise.' The Ephor laid both hands palms upwards on the edge of the table, and looked Vasilis in the eye. 'It did not have the desired effect. Quite the opposite. It was the action of a weak and fearful man, which I am not. I made a mistake. I will not make that mistake again.' Maria set between them a plate of nuts and cut-up cucumber and a bottle of raki. Neither of the men moved. The Ephor went on, 'It did not please the general that his cousin was a robber. I sympathised, ashamed myself that the robber is a Greek. So ashamed that I have not told the English. I have also promised your cousin the general that I should not tell Costas Pavlidis, the Epimelete. Neither the Epimelete nor the English know that you robbed the tomb. Only your cousin the general and I know. And several others in the Ministry have their suspicions, of course.' With well-pretended detachment Koraes took a piece of cucumber and chewed it carefully.

Vasilis reached for the raki bottle and filled both the glasses. He gulped his down and pushed away the glass with the back of his small, dirty hand. Koraes frowned at what a small hand it was, interested that a tough character like this Vasilis should have such sensitive hands. With such hands he could have been a poet, a painter, a violinist. There was distress in that small hand, and humanity. Was he discontented? Behind the dark eyebrows and the ugly low hairline, was there a soul that longed for truth and beauty? Koraes grasped his glass, deciding he did not want Vasilis to go to prison. He wanted him to act honourably and continue a free man.

Koraes tapped his glass on the table and wished Vasilis good health. Vasilis refilled their glasses. The Ephor pulled a neat handkerchief out of his top pocket and blew his nose, distorting his orderly face for a moment, before he drank down the strong drink which he'd have preferred to leave. But a glimmer had appeared in Vasilis's dark dudgeon which encouraged the Ephor; he wanted now very much to win through. With the raki burning inside him, carelessly stuffing the handkerchief back in his pocket, he asked Vasilis which of the English he liked best.

Vasilis's mouth puckered in thought. He was no longer suspicious. He leaned on his elbow and rubbed his cheek, pausing theatrically before answering. The Ephor pushed back his chair and crossed his legs, in the meantime inspecting the back of the old iron stove. He ran his eye up the galvanised pipe. 'Does it work well?' he asked, interrupting Vasilis's pause.

Vasilis's glance missed the stove altogether, taking in the long, flowery skirts that had just come in off the street. He dismissed them with a jerk of his chin and settled his small bloodshot eyes on the handkerchief in the Ephor's pocket.

'*Kyrios* Bill is my favourite.'

'And *Kyrios Klystopha*?'

Vasilis shrugged. 'He's a strange man.'

'How strange?'

Vasilis tapped his temple. 'He's a little crazy,' he said. Koraes ran his fingers across the table as if he were playing a scale on the rough board. He was intrigued, but not sure he should go on. Should he encourage Vasilis to think he understood Christopher? If he let him enlarge on Christopher's craziness, there was no telling what outlandish opinions might not start to flourish. Koraes rose to his feet and slid his hands into his jacket pockets; he decided it was time to leave. 'I like you actually,' he told the café owner. He held out his hand, which Vasilis shook briefly, before uncoiling himself from the leg of his chair and moving out into the middle of the room. Vasilis followed the Ephor to his car which was parked down the road, just where the concrete dipped into the gravelly riverbed. Overhead streaks of pink coloured the cloudy dusk, which the Ephor stopped to admire, wondering what his last words to this Cretan tomb-robber should be. Had he managed to convert him? He'd avoided threats, and tried to be his friend. Had he made him ashamed? There was no telling. He sensed Vasilis waiting a few feet away, not breaking the silence. The shrewdness of the man was daunting. He was, as the English would say, a chip off the old block.

'I think,' said the Ephor, pulling open the car door, 'that you should return the antiquities not to *Kyrios* Bill but to *Kyrios Klystopha* who is in charge of excavating the tombs.' He climbed into his car, and started the engine, driving off quickly, with his window rolled up.

The Ephor had nearly reached Ierapetra when he stopped and got out. The rushes rustled in the dark, the road by now running along the beach. He should have stopped earlier perhaps, where he would have had a view. But it was too dark to see. The rushes were a barrier, their cold sound unpleasant. He left his car and walked up the road, feeling the night hide his agitation. It was a shame that the tomb had been robbed. When Christopher telephoned he had been astounded because he trusted Christopher more than he'd let him think. He had believed Christopher would prevent such a thing. He hadn't confided in him, afraid that a 'suspicious' Ministry might have fired Christopher's antipathy to the Colonels and made him jump to false conclusions. Nor was anyone to know that he himself had been threatened with prison, not even his wife. He wished his actions to appear to be his own. The English were free people. They had a nose for coercion of any sort and could be too quick with their sympathy. If he were pitied he would be trapped in the presumption that everything he did he was forced to do.

Who would he be then? A poor thing, a hunted, frightened man. He would not be that. He was a Greek, an archaeologist, a cultured man. He loved his country and believed in its destiny, a belief no Englishman could understand. To them Greece appeared a shabby and badly governed country. But that would change. Greece had a great past and even Vasilis Andonakis was a true Greek; a shrewd and independent man, no cringer, a man of spirit. If only he weren't a tomb-robber!

Sideris turned back towards his car. He would eat something in Ierapetra before crossing the isthmus. He felt cold but happier. It was lucky that his secretary Mina had discovered about Vasilis's uncle – she had discovered it from a cousin who had gone to school in Ierapetra with Vasilis's father and the uncle who went into the army. Sideris needed only once to mention Vasilis Andonakis and Mina came up with the uncle who was a general, bless her. She was his truest friend. Probably Pavlidis knew that Sideris faced prison. Mina did not know. But she knew what a snake old Demetriadis was.

Poor Christopher was afraid his workman did it. He was suffering, the poor man. He loved that old man from Knossos who'd worked for the English for so many years. If Sideris had run into Christopher he would have told him not to worry. He kept a lookout as he drove back into the village from the police station, but Christopher was nowhere to be seen. Perhaps it was fit punishment. He should not have allowed the robbery to happen.

Costas Pavlidis gladly joined the English again for dinner. He sat down in the chair Jack pulled out next to his. He was amused to see so many English eat together in this poor room so far from home. There was something pathetic about it, which intrigued him. He took off his dark glasses and drank to Jack's health. Koraes too could have been there, which he would have enjoyed since he was such an Anglophile. The director's wife served everyone huge helpings of liver and apple, Costas's second excellent English meal that day.

But Pavlidis was not surprised that Koraes had kept his visit so short. He collected the gold and rushed off because he was a frightened man. He wasn't a bad man either. Costas admired the Ephor's scholarship, was proud that a Greek was so eminent in the field . . . and longed for Koraes to have a higher opinion of him. But the Ephor made plain his low assessment of Costas's worth. He refused ever to hear Costas's theories on Minoan religion, or trade in faience, both topics that Koraes had written about. Even the Englishman Christopher had shown more interest in Costas's ideas that morning at the pot shed.

'Have you seen the schoolmaster?' Christopher was asking, sitting opposite. He looked cold and tired. The cook Margaret brought out other dishes of bean salad and cheese, but no one had appetite after so much liver. Except Christopher who had tucked in, his thin jaws chomping up and down like a dog. Costas also noticed that Christopher had drunk three glasses of wine, and was in the act of pouring out a fourth, offering Costas more wine before setting down the jug. Always courteous, this Christopher. When the director down the table had complained about the Ephor, Christopher had not ignored Bill as the others did but stopped eating to shake his head in sympathy. He was such a puzzle. Cold and tired and hungry, courteous and perhaps an alcoholic, Christopher was not what Costas would have expected of such a well-known English scholar. He was too natural, too polite, and not aggressive enough. Was he hiding something? Jack described him as a 'typical aristocrat'. Jack disliked him, which made the Epimelete suspicious.

'I'd go and have a talk with him, if I were you,' Christopher advised.

'Does he still think the German did it?' Costas asked. Christopher would not answer.

The Epimelete dallied over his liver. Christopher and Jack finished and left. The beautiful Mary Elizabeth sitting next to Bill carried her plate out to the kitchen. Costas moved to her seat, pretending not to have realised it was hers when she came back through the room. She said good night. The room was finally empty. Jenny and Laura in the kitchen behind the partition clattered the plates. Bill was tapping ash into discarded bits of liver. He said nothing when Costas sat down. His mind was on something else.

The Epimelete shoved his hands in his pockets and fingered his car keys. 'I hope my presence does not make it more difficult for you,' he apologised. He cleared his throat. 'Am I in the way?'

Bill reached for the wine jug in a distracted way. 'Where's your glass?'

'No, thank you.'

'You're probably right.' Bill refilled his. 'That policeman. Have you seen him? How's he getting on?'

The Epimelete laughed deprecatingly.

'He's doing his job,' the director persisted. 'I'm all for it.'

'He has no idea what he's doing. He's an ignorant peasant,' Costas snapped.

'A case of robbery is a job for the police, surely.'

'But of antiquities,' Costas argued. 'What does that policeman know about antiquities? Whoever robbed the tomb knew more than that policeman knows.'

The director's eyes fell to his messy plate. 'What does the Ephor think?' he asked. He was turning his glass uneasily. They watched the wine in the glass remain still.

Costas shrugged and pursed his lips. He sighed. 'The Ephor is a very strange man.'

'I'm sorry to have missed him.'

An uncomfortable silence ensued. No sound came even from the kitchen. Costas reached across to turn off the lamp where the flame was so low it was pointless. 'Do you mind if I ask you a question?' he hurried on. 'You won't be offended?'

Bill blew out smoke from his cigarette.

'Why did Christopher order the workman Andonis to return? Today he telephoned his wife to make him come back.'

Bill hid his surprise behind another puff of smoke. 'He's very fond of him.'

'But he was guarding the tomb, was he not?'

'Who?' Bill asked sharply, on his guard now. Costas waved his hand to mitigate any offence.

'I hope you don't think I'm prying. But it was this same Andonis whom Christopher has ordered back who was the night watchman Monday night, was it not?'

Bill's chair scraped the floor as he pushed it back from the table. He stretched out his legs and rummaged in his trouser pocket for matches.

'It is not easy for me to ask you this, but it is puzzling,' Costas persevered. 'Christopher's behaviour is very puzzling. Is he a rich man?' he asked.

Bill shook the matches. 'Who are you talking about?'

'Christopher.'

'Rich enough.' Bill lit another cigarette.

'He's a very nice man,' Costas was fingering his car keys again. 'And a great scholar. I've read his book on palace architecture which is extremely good.'

Bill tapped ash on to his liver leavings and tapped the cigarette again on the edge of his plate before he spoke. 'What are you wanting to ask me?'

'Why do you think Christopher wanted Andonis back? Does the workman know something about Christopher that Christopher is afraid he will tell people if he does not keep an eye on him?'

Bill exploded. 'You are suggesting Christopher engineered the robbery. I'm astounded.' Both his fists hit the table. 'Never in my life!'

'Please, please, please . . .' Costas held up both hands. 'It's only I don't understand,' he pleaded. 'And it is strange Christopher did not tell the Ephor that there was a night watchman, because obviously the night watchman helped rob the tomb. An accomplice at least.' Costas was frightened. Bill had jumped to his feet, his anger skimming the room as he rushed with his plate to the kitchen.

'What do you mean, Christopher didn't tell the Ephor?' he shouted over his shoulder. 'I heard him myself tell the Ephor that Markakis was on guard and disappeared because he wanted to see his son in hospital. He and I want the workman back, because he's a good workman.' Pavlidis could not make Bill turn around. He addressed him across the room. 'I'm going to bed,' Bill bellowed back as he dived into the dark beyond the partition. 'If there's anything else you want to ask, it must wait until morning.'

Bill found Jenny and Laura huddled together just inside the door, Jenny in her rubber gloves, Laura holding a pan and a dishcloth. 'You two look ridiculous,' Bill hissed, plonking his plate down on the table behind. Both women pressed fingers to their lips until they heard the street door rattle shut. Laura peeped around. The Epimelete had gone.

'But how could he?' exclaimed Laura.

'Let's get these dishes done.' Bill was at the sink, pouring water from the kettle into the plastic bowl.

Jenny hooked the Camping Gaz onto a higher nail and turned it up so that the sizzle and glare pressed them to work.

Laura was the least preoccupied. In England it was always like this: Bill washing, she drying, Jenny putting away – the knives over the Aga, the pans on the shelf above. Everything in Jenny's cupboards and drawers was in such a muddle that only Jenny could put away. Jenny's kitchen was never as neat as Laura's. Or as clean. Her shelves were gritty with spilt sugar and herbs, butcher's tickets left lying in flour dust. But how warm it was! Laura loved the ease and security of her friends at their Aga hearth where she could dream that if she had an Aga she'd be just as happy. It was different from this dark hole where the plates and glasses clattered when you knocked against a trestle table.

'Just like old times,' Laura hummed, rubbing her thigh where she'd bumped herself.

Her amiability misfired. Jenny was so befuddled from her walk with Adam that the Epimelete just now was a relief, restoring her to a serious present which Laura had no business to ignore. Bill jabbed at her with a pot, pushing it into her hand.

'*Did* Christopher telephone Andonis to come back?' Jenny asked. Bill sloshed water and clanked plates, the noise he made sounding futile in the primitive kitchen. 'I thought they would have arrested him. He must have had something to do with it. Adam thinks he's been framed.'

Bill swirled round, spraying them with soapy water. 'Adam's a damn fool,' he shouted.

'Bill!'

In a sulk, Bill turned back to his plates in the plastic bowl. But Jenny had stopped putting away. What had he done? Couldn't Jenny see that his outburst was because of the robbery? Laura had to hurry to keep up. She tugged at Jenny's sleeve. 'Come on. Buck up, Jenny darling.' She pleaded with the still figure leaning against the trestle table.

Jenny dropped the lid on to the floor and walked out. The scrape and rattle of the street door died behind her, leaving headquarters to Laura and Bill.

It was across the street from the mayor's new café, where the light from its large, clean window picked out the gold ring on Christopher's finger and his thick-soled brogues, that the Epimelete saw an old Cretan in a white shirt talk with him. The worn voice of the old man muttered in such colloquial Greek that Pavlidis made nothing of it. He pressed himself against the wall at the side of the café by the door and shut his eyes, straining to hear. He had to blot out the shouts of the mayor and the policeman inside who were playing cards.

'But he'll be all right,' said Christopher. Costas heard that. Then more from the old man who must be Markakis. He would have arrived by now; he was telling Christopher everything.

'Good . . . Right . . . Well then . . . Quite . . . So . . .' So far Christopher's talk was patter.

Inside the mayor slapped down his cards and bellowed. Costas prayed for him to be quieter, certain that at that very moment he'd missed the crucial word.

Christopher was moving on. They'd finished. What promises, threat, bribe? What had he missed? Those brogues crunched the road.

'Tomorrow then. Good night,' Christopher called. No answer from the workman. The night lost both men, the Epimelete left fuming in his hiding place.

Christopher had not found Andonis's return a consolation. He was relieved that his son was going to be all right. It was a horrible thing, to be hit on the head by drunken passengers, a motiveless, indecent violence that had shocked the father. It would take time for Andonis to recover from that. But also the night of the robbery sat indigestibly between them. Christopher did not want to ask Andonis anything that would upset him, and he was afraid. Had he helped rob the tomb on that night when he was so worried and so mishandled by Bill? Christopher did not want to know. Stubbornly Christopher would hold on to his hope that Andonis knew as little as the rest of them.

Bill, Christopher knew, was fed up with him. Christopher sympathised. Bill hated the shame this robbery brought them in the eyes of their Greek colleagues. It was a huge pity. Should Christopher try to make up for it? Should he get on with the new trench, and if it turned out to be unrobbed, guard it himself? He was a fool to imagine that the young Epimelete had the authority to take over. He was too young and inexperienced although an intelligent chap. He shouldn't mind anyway if the Epimelete wanted to help.

Right. Christopher yawned as he hung his trousers on their nail. He rubbed his eyes, standing bare-legged in the sleepy light from the lamp which he'd set on the ground by the bed. He was too tired to read. He burrowed into his sleeping bag. He pulled the nylon over his shoulders, resting his chin on the edge. He'd decided. Tomorrow he'd get on with the new trench, and please God may it be a good one, he prayed, shutting his eyes.

But Costas Pavlidis was too cross to sleep. His hatred of the Ephor grew by the minute. Koraes had pretended to him that he had no idea who had done the robbery as he stood smoking a cigar, with his other hand resting on a stack of books like a *pasha*. He'd cheated him. Of course Costas in his ignorance would suspect one of the English. Was that the Ephor's plan?

Inside the mayor's café the Epimelete had wanted to sit alone and think. But the schoolmaster was at the table next to where they were playing cards. Jack had pointed him out earlier, and said he only came to this café because it was clean. Costas was obliged to introduce himself although he was shy, as all Greeks are a little shy of the village schoolmaster. His schoolmaster in the Peloponnese hadf been a good man, but strict. Costas could imagine how that walking stick hooked over the back of his chair was used during the day.

'I know who you are,' said the schoolmaster, before the Epimelete spoke.

Costas offered the schoolmaster a beer, which the schoolmaster accepted.

'This robbery is a bad thing. It gives the village a bad name. That's why you are here,' he continued.

The two men watched the game of cards, Costas unsure what to say. The very fact that Christopher had asked him several times if he'd yet seen the schoolmaster made Costas even more shy. And suspicious. Why was the Englishman so eager? If Christopher had not mentioned the schoolmaster, the Epimelete would have gone to him straight away, knowing he would have ideas. Now, however, he wasn't sure he wanted to hear those ideas. He had his own ideas.

'Where are you from?' asked the schoolmaster, when the beer had come and they'd drunk to each other's health. But the schoolmaster did not know the Peloponnese. His family, he told the Epimelete, were from a village up in the mountains which had been destroyed by the Germans. It no longer existed. He'd gone to the gymnasion in

Ierapetra. For twenty years he'd been the schoolmaster in this village.

He was an archaeologist. He knew all the sites in the region. He had his own small collection of finds which he would gladly show the Epimelete if he'd like to see it. He had shown it to the English archaeologists who were impressed. They were his friends. Costas listened meekly, sipping his beer.

'You have many visitors in the village, I understand. Is that good?' Costas asked, scanning the room when the schoolmaster had come to a halt.

'No.'

'They are all hippies, you mean?'

'Germans.'

'I heard about the frog incident from one of the English women.'

The schoolmaster wagged his finger, Costas's dark glasses hiding his amusement. The schoolmaster's chin jutted out at Costas like a battering ram. 'He's your robber! They always take. They take, take, take. To fill their museums in Germany, they rob us of our past!'

Costas took off his glasses and leaned his head on his hand. It was what he'd often said himself when he wanted to sound off to his friends in Athens about the wrong of foreigners having the right to excavate in Greece. But he didn't believe it. Foreigners were needed, unfortunately. The Germans worked hard and were brilliant scholars. Costas couldn't agree. Yet such patriotic resentment was not wrong.

The Epimelete blinked away tears. He yawned, his leather jacket squeaking like a rusty hinge when he covered his mouth with his hand. He pushed back the hair from his face. 'To insult you with a frog does not mean the German's a tomb robber. It's one of the workmen,' he muttered, eager now for bed.

'The redhead. I know.' The schoolmaster's finger stabbed the air and nailed the man. Costas was struck. He leaned forward.

'Why?' he asked.

'He's German. His father was a Nazi. I'm right. You'll see.'

The Epimelete slumped back into his chair.

Half an hour later the policeman from Ierapetra, Sotiris Georgakis, saw the Epimelete sitting in the mayor's café in his leather jacket. Georgakis turned heel although the Epimelete had his mouth open and noticed nothing.

CHAPTER 26

Laura was surprised how she longed already for the end of the excavation as she shivered in the cold, dark room, groping for something clean enough to put on. Such a sudden, phoney desire to go home startled her. London would be no fun. The leak in her roof needed repairing and she dreaded not finding someone to do it. The fussy façade of her little house with its silly stuccoed pediments of grapes half way up the window frames and 'Carlyle' over her doorway had no appeal. The Indians' shop at the end of the street was as bad with its dusty cauliflowers on the doorstep and the assortment inside of rice, Polo mints, tins of corn – its fusty, sweet smell as horrible as this rubbishy room smelling of urine. Annabel grunted in the gloom as she pulled on her riding breeches. Laura blanched.

It didn't matter that her pink T-shirt was dirty since she'd be back sieving, she supposed. Laura bound up her hair in the first scarf she could yank out of the heap in her suitcase and limped to the door, desperate for the loo. She'd have some way to hobble before she could reach blissful relief, in too great an agony in the meantime to appreciate the clear, cold day outside.

A little later Laura was on her slow way to the site. There was the first sprinkling of snow on the mountains, which showed faintly white in the dawn light. Not wintry but incredible seen from the valley where not even rain had laid the dust. But if she'd known when Michaelis was going on about the wonderful nuts and honey up in those mountains, that it was snowing, she'd have refused to go back with him to his room. And certainly she would have said no to the proposed trip on Saturday. Snow put her off everything.

Nonsense! Laura made herself admire the pale peaks of Mount Dikte. She upbraided herself for shying away from real adventure. She loved adventure. Michaelis was adventure, the snows were adventure, this dusty valley was adventure. Forget Bill and Jenny and Annabel. Who in London wouldn't envy her? Compare herself with her idea of a dull commuter chewing a Mars bar on the District and Circle line, looking forward to *News at Ten.*

When she reached the top of the hill, Laura went straight to her stone seat by the two piles of earth. Susan was already shovelling unseived earth into buckets; Laura's sieve was leaning against one of the full buckets. Also there watching was the Epimelete. Laura spread her knees and picked up the sieve, remembering that Michaelis said that the Epimelete was a spy. A second later she found that the Epimelete stood behind her with a full bucket; he asked if she'd like him to dump the earth into her sieve. She could make out the back of Michaelis's head 100 yards away, but the rest of him was hidden by the hill.

Laura assumed the Epimelete meant to flirt. She waved him on, giving Susan a cautious glance. Susan gave Laura a benevolent look back which amused Laura who thought what a silly girl she was.

Costas's eyes were on Laura's narrow thighs supporting the crude sieve. What a beautiful middle-aged woman, he thought, but probably mad since she was willing to work on a barren hill sieving rough clods of earth like a peasant, which would ruin her hands. It concerned him to watch her so misuse her beauty, marvelling at the English propensity to behave unnaturally. How it disgusted him when his mother did rough work like this, deploring the poverty that forced her to have cracked hands and bunioned feet! Whenever she groaned that her feet hurt Costas would retreat, squeamish as a boy. His squeamishness now seemed a natural, educated distaste for the crude and crippled, which was a characteristic widely shared in the civilised world. Why this English woman who was not poor chose to work herself like this puzzled him. Was she too afraid of something to say no? Or was she a saint? Costas caught the corner of a smile coming round her face. She was no saint.

'I met your schoolmaster,' he informed her.

'Why mine?' Laura asked, startled.

'You said yesterday about the frog and the German tourist. I remember.'

'What a good memory!'

'How could I forget?'

Laura sighed, accepting the rigours of a new flirtation – which if she'd not felt so wobbly she'd have enjoyed, despite this Epimelete's fat tummy. And Michaelis's suspicions. But she wasn't herself. Last night

Jenny had upset her when she marched out on Bill, and Bill nearly hit her with a pan. It was Bill's fault that she woke up this morning wanting to go home.

'I'm flattered,' retorted Laura as she waved on his dumping of buckets.

'You hurt your hands,' he chided her.

'When is your birthday?' Susan interrupted. She stood idly, with hands on hips, waiting for the Epimelete and Laura to catch up. She puzzled the Epimelete. He set down his bucket and rested his hands on his hips, the other side of the sieved heap of earth. Christopher, as he approached them from the new trench down the hill, saw their opposing stances and was dismayed. He wanted no arguments, especially with the Epimelete. He waved and halloed.

Laura waved back.

'Why do you ask me that?' asked the Epimelete.

'Because I can tell you about yourself, if you like,' Susan volunteered, swinging both arms. Costas threw back his head and laughed, which relieved Christopher from a few yards away.

'You're an astrologist!' Costas waved his dark glasses at Susan. 'I should have known.'

'She tells me I'm on the brink of a great change, and I don't like it,' Laura joined in, her eyes on Christopher's tanned face and rolled-up sleeves coming towards them. He was at last in a clean denim work shirt like the one he'd worn at the beginning.

Costas had moved to a few feet from the far side of the unsieved heap when Christopher reached them, his dark glasses back on his nose. The hard finish of his leather jacket helped him dissemble his confusion in Christopher's presence. It was only yesterday that he and Christopher had had such an interesting discussion on faience, before the business of the night watchman had given him the wrong idea.

Laura sensed the Epimelete about to hiss from behind his leather jacket like a cornered animal. She forced with her eyes and smile the young Epimelete to join in her banter with Christopher. 'I advised him not to tell Susan his birthday. Or else. He's in for it, *n'est-ce pas*?'

Andonis was very close behind Christopher and limping a little. She called to him. She waved.

'A coffee?' Christopher asked the Epimelete.

'We're both limping!' Laura exclaimed. She picked up her fig stick and held it out to the old workman. 'Use my stick.' But Andonis's eyes were fixed on the ground. He looked older. His shoulders were hunched, his pale blue shirt fixed to him in inert creases.

Costas accepted Christopher's offer of coffee, and Andonis followed behind as the two men made their way over to the olive tree.

Christopher did not want anyone but Andonis to be present when he informed the Epimelete that they were uncovering a second, unrobbed tomb. Christopher spooned the Nescafé powder into the shaker, tilted the water jar until enough had gurgled out, screwed on the top and shook until there was a *frappé* to pour into three plastic beakers. Andonis's presence made Costas nervous, but he was too in awe of this much older, senior English colleague to speak out although he was now convinced that Andonis was the robber. The guilt in the workman's bent shoulders and cowering behaviour proved his hunch that the Englishman was, perhaps innocently, protecting the culprit.

This so preoccupied Costas that he did not guess at Christopher's reasons for taking him away from the others to tell him about the tomb; nor did he realise that Christopher expected a reaction, as he sipped the Nescafé *frappé* in silence. Christopher expected the Epimelete to blow up with excitement and insist on supervising the new tomb. Christopher waited. Andonis had squatted down on his haunches to draw a long-legged hare in the dust with an olive twig.

'Good,' pronounced the young Epimelete without enthusiasm. 'My congratulations.'

Christopher cleared his throat and collected the empty beakers. He leaned them against the tree in a stack beside the plastic shaker. The tubby Greek in the leather jacket and the tall Englishman sweaterless in his khaki trousers had nothing more to say. Christopher hurried back to the new tomb, assuming that the Epimelete would come if he wished.

It was soon after that that Costas, weighed down by his mission to prove Andonis guilty, walking round and round the edge of the emptied tomb, spotted down in a corner of the trench under a pile of dried thistles a turquoise something. He jumped down and pulled at it, picking the thistles off a scarf with black lines on a turquoise ground. He crumpled it in his hand. Its strange soft feel convinced him that it had been there since the robbery. It was blown deep into a corner because two days had elapsed. At night the robber would not have seen it blow away either, although it was far too bright to be missed if it had slipped off during the day.

Costas stood with his back to the women; he faced the pale peak of Mount Dikte and wondered what his next move should be. How would Andonis have come by something so fine? It said 'Harrods'. Ah! It was a gift from Christopher to the old workman. An odd thing to give a man. Perhaps Christopher gave it to his daughter. Costas glared at the bold black lines. He pulled it taut. What a frustrating find!

'Come and have a look!' An English shout. Costas looked round. Laura and Susan were leaving their work. The dark-haired boy called Edward had called excitedly to the women to hurry over. Laura was limping on her stick. Slowly the Epimelete followed, stuffing the scarf in his pocket. In front of him was the redhead the schoolmaster said was the son of a Nazi. The redhead stopped to take the arm of the beautiful Laura.

Christopher laid in the Epimelete's hand an Egyptian scarab. Manolis was on his knees picking round a jumble of bones, three skulls, a stone cup, pots and a cache of seals of which only the edges showed, like dominoes in the darker earth. The buff scarab had lain apart, its special preciousness preserved for thousands of years; it was the first thing found in the debris of white *kouskouras* from the collapsed roof of the tomb.

'Isn't that terrific?' Christopher leaned over Costas's hand, gingerly fingering the scarab. He turned it over so that the Epimelete would see the hieroglyph on the other side. The Englishman was breathing hard. Pleased by such excitement, Costas calmly handed the scarab back to Christopher and moved around so that he could see better. It was amazing! Costas forgot about the handkerchief, down on his knees to peer over the side.

It was Laura who brought him back to the misery of the other, robbed tomb. In her excitement she reminded the Epimelete that he had also found something. 'What was it?' she asked, standing opposite beside the redhead. It was because this new wonderful discovery had disarmed them all and inspired harmony, as they stood or sat or crouched in a circle looking down, that Laura had dared ask. He'd not volunteered to show it, and he was a Greek official with the authority to do as he liked. Everything there was ultimately his business, belonging to his country. Could the Englishwoman see how impudent she might appear, asking him such a baldly inquisitive question?

The Epimelete stood up and brushed himself off. He looked round for Andonis. The group broke up; Christopher urged Susan and Laura to finish sieving quickly so that they could help him. Edward and the redhead had gone back to their other tomb. Only Laura still lingered.

Costas walked over to where Andonis leaned on his shovel smoking a cigarette. He handed him the turquoise scarf.

'What is it?' asked Laura. She limped up from behind to see. 'Show me,' she demanded.

With a shrug Andonis handed Laura the scarf; the Epimelete had scared him badly, but the scarf meant nothing. He'd never seen it before, though he was shaken by the Epimelete's strange action.

She'd never liked the scarf. It was her first thought, expecting to see a gold pendant, or a seal, or something like that, something ancient and beautiful that the robbers had overlooked in their rush – just as she had found, sieving, those two gold pins. She'd watched the tubby Epimelete jump down into the empty trench and seconds later climb back out clutching something. He'd not wanted to show her and Susan what it was, turning his back on them as he examined it. But her scarf! What could be valuable or important or even interesting about it? John had given it to her last Christmas, and she'd pretended she liked it, bringing it with her out here because Jenny had said bring old clothes that it didn't matter ruining or losing.

She'd given it to Michaelis. Now she remembered, blushing as she recalled the circumstances.

Costas noticed how Laura hesitated and blushed, before she claimed the turquoise scarf. She thanked him, but fingered it suspiciously.

'It blew off my head,' she said.

'When?' asked the Epimelete.

Again Laura hesitated; she folded the scarf into a smaller, smaller square. 'Yesterday. I'd put it on for sieving.'

The Epimelete left her. Laura watched him go in a huff, as if she'd said something that offended him. He tripped. Laura giggled, his sleek jacket absurd as it stumbled out of sight.

But she sobered instantly at the sight of Andonis, who stood by his shovel shivering, his still face blind from fright. Laura looked around for Christopher. Where was he? Andonis needed him.

Christopher was on his knees inside the trench, brushing alongside Manolis. He stopped to riffle through his notebook, propping it on his knee to write. Laura moved as near as she could and asked in a low voice if she could speak to him a moment. Christopher waved her away. He shouted to Andonis to fetch the ladder. The old workman dropped his shovel and ran. Christopher pulled his camera strap over his head, ignoring Laura.

Laura decided that was a good thing as she hobbled up the hill. Edward laughed when she passed by his trench. Michaelis was picking and singing and shovelling like a madman. Michaelis shouted at Edward to 'hurry, hurry, hurry', and Edward nearly fell in when his foot caught on a root. 'I can't keep up with him,' he shrieked, grinning at Laura, his young face smeared with sweat and dust. Michaelis let out a whoop and wheeled the pick over his head. Edward was delighted.

'He'll break anything there,' Laura called, but Edward wouldn't listen. No doubt she appeared to both in as silly a huff as the Epimelete, as she limped back to her sieve. Michaelis would be very surprised when he found out how angry she was. When he saw how even an Englishwoman could lose her cool, he'd be sorry.

He'd be sorry. Oh, how he'd be sorry! He'd be so damn sorry, Laura muttered, as she pushed through the last bucket of earth. Her sullen silence unnerved Susan. She'd make him sorry, Laura promised herself hopelessly.

There was no chance until after lunch to have her say. Laura wavered. Was she wrong to think the Epimelete meant to incriminate Andonis? Andonis looked so frightened. But perhaps it was on Tuesday as she'd told the Epimelete, and not on Monday night that the scarf was blown down into the empty tomb. She hoped so. If only Michaelis weren't the robber. It would be so embarrassing.

This Thursday morning of the third week of the excavation became the longest morning Laura had had yet on this adventure of hers. Her aggrieved impatience, like a wind blowing every which way, ripped memories from context and longings from nowhere; regret and anger and dismal dejection made the minutes creep by. Her guilt was the dismal part, when Susan saw Laura look most sullen; her hands moved back and forth unwillingly, her head right down. It was so familiar to hate herself, and so bleak.

Christopher came over to inspect. One heap of fine, sieved earth testified to two and a half days of work. Christopher said the two gold pins were worth it. Susan and Laura were to come and help him now with the new tomb, which was a huge job.

Christopher would not stop. He asked Susan to stay with him. Laura, he thought, looked tired and in pain. He noticed that she showed no excitement. Monday she'd been so different, when she'd packed the first finds from the robbed tomb into zembils. He missed her enthusiasm.

When Laura limped through the door of headquarters, the first to arrive, and alone, Jenny ran up to apologise for walking out on her in the middle of washing up. 'I couldn't help it. How's your ankle?' Jenny looked cold despite the beige cardigan she wore over her white polo-neck; she was rubbing her arms which were crossed in front of her. Laura managed a smile and followed Jenny into the kitchen; Jenny had even put on socks and an old brown wool skirt. 'Is it hurting?' Jenny persisted. Laura described the cache of seals and the scarab that had already been found in the new tomb, and carried the chicken stew back into the front room. She sat next to Adam and watched him finish off three helpings.

Her bandaged ankle and the stick made Laura all the more conspicuous as she hobbled down the village street. The men watched, amused, Laura sensed, to see the beautiful Englishwoman so crippled. It made it difficult for Laura to find Michaelis without being noticed. When she passed the bakery *Kyria* Georgia rushed out into the street to ask Laura if the ankle still hurt.

'You use the stick!' exclaimed the kind woman who'd rescued her on Tuesday. 'My son made it for you.'

Laura was overcome. *Kyria* Georgia was delighted.

'He's a good boy,' the woman told her, squeezing Laura's arm. Near tears, Laura moved on.

But Michaelis was not in his room. The peeling green door was ajar; inside, the iron bed was just visible in the gloom. She backed off before anyone could see her, and sat down further on to look out at the sea.

Maria, Vasilis's wife, waved to Laura through the dirty café window; Laura spotted Michaelis inside sitting with Vasilis. Instantly Vasilis was at the door urging Laura to have a drink. She couldn't refuse. As Vasilis pulled up a chair, Michaelis shovelled more lunch into his mouth. Maria came over to ask if her ankle was painful. Michaelis chewed his food like a contented bull. He wouldn't give Laura even a smile. Why was he so unfriendly when among his fellow Greeks? Was he ashamed of her? How pathetic of her to think that. It showed Laura how low she was. She must pull herself together. She was spurred by fresh anger. She told Michaelis that Bill wanted to see him, and watched his face change, his chewing stop, his fork go down. She added that the director was waiting for him at headquarters right now.

Michaelis jumped about when they turned off into the field behind the priest's house where three goats were gnashing through the thistles. He thought her lie funny. Laura complained that it wasn't that funny, wishing Michaelis would stop laughing.

'You're coming to the mountains with me. Agreed?' he challenged her.

Later he was hunched on the iron bed with arms dangling between his legs, in a dejection as off-putting as his high spirits had been in the priest's field the hour before.

'But have you thought in what kind of position this puts me?' Laura asked. She was still in her dusty T-shirt and jeans, swishing the turquoise scarf in front of Michaelis's bowed head. The longer they talked the angrier she became with this wretched Greek. He'd betrayed them all. Hadn't it occurred to him that they trusted him?

'I don't know what "trust" means,' muttered Michaelis, feeling as sorry for himself as Laura was feeling sorry for herself. 'You won't like me any more. You won't come to the mountains.' He raised his big head and jabbed it at her like an armless Punch. 'Last time you didn't come with me and I robbed your tomb. You, what is the word . . . "betrayed" me!' It was his bushy eyebrows bunched above his big eyes that made him look so fierce.

'Don't be ridiculous.'

'You laughed at me, after you'd given yourself to me. You'd lain with me. You were my woman.'

'How biblical you sound.' But Laura did not laugh or snigger or even sneer; her tone was low and apologetic. She was afraid. She took a few steps away from the bed, folding up the scarf.

'I blame you,' he whined.

With clenched fists and bared teeth Laura screamed, 'Don't ever say that again!' Michaelis's mouth opened in astonishment. Then, as unexpectedly, she burst into tears. Not running away, not throwing herself at him, not supporting herself on the door, Laura wept standing in the middle of the floor like a statue. Michaelis sensed she was ashamed from how she spread both hands across her face.

Michaelis had never seen a woman be ashamed of weeping. It was unnatural, and he pitied her. He loved her. She was so beautiful . . . Gently he guided her to the bed and sat her down. He pushed her thick black curls out of her face and kissed her ear. He apologised and gradually Laura calmed down. She let him take her hand.

It was strange that he could speak English although he had known so few English people. Several times Laura had remarked on this but then forgotten to ask Michaelis why; she asked him now and he told her his father had learned English in the war. His father had been a runner for the English, and there had been one Englishman in particular whom his father loved. This Englishman called John had had long conversations with his father, explaining to his father the English system of government and reciting to him his favourite English poems. He taught Michaelis's father many English words, and gave him a book called *The Wind in the Willows*. It had been John's book from England and Michaelis's father kept it in the cupboard in the front room. Michaelis knew the pictures well, of a mole and a rat behaving like human beings with caps on their heads. His father had urged him to learn English in school so that he could read it. But it was from a Dutchman that Michaelis learned most of his English; the Dutchman had employed him to help in his new plastic greenhouses and Michaelis had insisted on speaking English although the Dutchman wanted to speak Greek. He supposed he might know enough English now to read *The Wind in the Willows*. He knew the title. He'd known the title from when he was a child, from his father.

'*The Wind in the Willows*,' Michaelis sighed inconsequentially as he sat beside Laura on the edge of the bed in the windowless room, holding her hand while a thin slice of bare light coming through the warped door revealed the dirt floor and the black leg of the bed. At the same moment that Michaelis was curious to know if Laura had ever read the book, Laura, dumbfounded, was asking Michaelis to repeat what he'd just said. She was so moved and excited when he repeated it all, asking him wistfully several times why he had not told her before, that he began to wonder at himself.

'But it explains so much,' Laura kept saying, 'about you.'

'What?'

Laura shook her head and held up both hands, shrugging her thin wide shoulders at the obviousness, the obviousness.

CHAPTER 27

Before everyone at dinner, standing in the kitchen door with a bowl of half-finished sherry trifle held carefully in front of her, Jenny announced that she was off to the mountains that weekend and would be taking the car.

Heads turned. No one was thinking about the weekend yet, a whole day and a half's work still to be done before then. Everyone looked at Bill who was still eating, but he wasn't listening. He'd already informed everyone that he'd be taking food up to Christopher as soon as he'd finished his trifle.

'It'll be jolly cold,' Laura piped up. She felt eyes on her now and saw Adam foolishly smile at her before he switched his attention to Jenny who still blocked the doorway.

'Can I come with you?' Suddenly Adam was asking Jenny; a sheepish, insinuating triumph lit up his thin face. Laura forgot to chew her next mouthful of trifle. Without answering, Jenny disappeared into the kitchen.

Laura sneaked a look down the table at Bill who was scraping his bowl, his confident head and shoulders hunched over the table. Laura was awed and thrilled. So blind he was! And so compulsive and arid these days, like an offensive mother who bossed her child about dreary minutiae of clean hands and unmade bed until the child needs to find someone else to show real interest.

But could Jenny be letting loose with Adam? Could Adam show enough interest? Wasn't Jenny too old?

Laura hurried with her plate into the kitchen. Jenny was scowling at the trifle bowl in the sizzle of the Camping Gaz which shone down from its nail on to the globs of grapes and apples and jammy cake in yellow custard. Laura peered at the mushy leftover which so preoccupied her old friend. But there was not a glimmer of a response. Jenny went on frowning furiously at the bowl. 'What do you plan to do with this?' Jenny rapped out, knocking into Laura as she whirled round on Margaret who'd just walked in with a stack of dirty dishes.

'Adam had two helpings!' Margaret defended herself, dumping the plates noisily into the plastic washing bowl.

'What do we do with the rest of it? There's masses left.'

'Give it to the chickens.'

'Whose chickens?' Jenny looked so worried that Laura gave her shoulders a squeeze.

'There's no shortage of chickens in this place, darling,' Laura coaxed, removing her arm when the shoulders remained rigid. Laura left, ruffled. She found herself at the far end of the room where Bill still sat over his empty bowl. The hubbub of scraping chairs and tinkling glasses as everyone cleared their place was in careless contrast to Jenny's stubborn gloom back in the kitchen. Laura decided to tackle Bill and make mischief, to forget her own predicament for a moment.

'Aren't you going to the mountains?' she asked. She tilted her head, running her tongue coyly along her lip.

Bill pushed away his empty plate which Mary Elizabeth scooped up and slid on to hers. Bill protected his glass with both hands. 'I must see how Christopher's getting on,' he moaned, gloom puckering his face in the faint light. Had she caught a gleam of deep upset? It might, of course, have been a trick of the sputtering lamp, which she turned off since the flame had sunk to a thin blue line which gave no light at all. She and Bill were now quite alone, the toing and froing of the others as uninterfering as an empty room.

'Jenny's in a bad way. What's wrong?'

Bill wouldn't tell Laura that he knew now who'd passed him in the dark coming up from the beach, after he'd had his talk with Christopher. At first he'd thought they were a couple of hippies. But later it came to him that at least one of them wasn't. He'd always relished the way Jenny hitched in extra breath when she laughed. That hiccup of delight gave his wife away, hidden under the rug with someone he didn't know. He could do nothing. What could he do?

Bill shrugged. 'I have too much else to think about.' Fortunately Bill could not see Laura blush as she pretended she was simply a concerned friend. If he knew she was the confidante of one of the robbers he'd go mad. She would never be invited down to Little Compton ever again. Nor would Jenny ever forgive her since she'd betrayed her husband. Laura took a sip of Bill's wine . . . And poor Christopher was keeping awake all night in case the robbers tried to rob them again. She could save Christopher that if she confessed. But he also would despise her. She needed time.

'I don't know what you mean.' Bill was glad Laura had come to sit down next to him. She was an old friend of them both.

'Nor do I, really.'

Jenny barked at Bill on her way through that Christopher's food was waiting in the kitchen. Laura ducked, avoiding the embarrassment of Jenny's temper. Jenny banged the door shut behind her. Bill rose to his feet.

'Coming?'

Laura winced. 'Where?'

'You might as well. We'll take the car.'

A picture of Christopher guarding the tomb, hunched over a fire in the wild Cretan night, titillated her lust for adventure. Go with Bill and tell them both now that Michaelis was her lover and one of the robbers. It would be the right and brave thing to do.

But, oh dear. Laura pushed the hair out of her face and stood up. 'I don't think I will, if you don't mind. I'm awfully tired.' She was suddenly exhausted, shivering. Bill had disappeared into the kitchen. He was coming back carrying a plastic bag.

'Sure?'

Laura clutched herself with crossed arms.

'I hope there's a torch in the car.'

Laura followed Bill out of the door. When they'd reached the car and she'd found the torch in the glove compartment she was still insisting that she'd best not go, determined not to. But Bill suddenly put his arm around her and hugged her.

Jenny had given their secret away like a madwoman. Adam wondered, walking back to his room, if he minded. His success with Jenny was unexpected and perplexing. She'd allowed his hands to wander all over her the evening before; she fell back on him in a swoon when he fingered her nipples through the white sweater. Whenever had Bill last made love to her?

Which was sweet. Adam didn't mind that Jenny was so desperate for sex. Her feelings needed to come out in a great whoosh. This surprisingly didn't disgust him. His squeamishness, he was finding, didn't extend to feelings, even when those feelings were centred on him. He liked to see his mother's middle-aged friend suddenly push back the curtains and show her need for sympathy. It was exhilarating; he was discovering in himself strengths he'd not known he had. If she'd let him, he'd make love to her, and give her lessons in drawing if that would please her. Whatever she wanted from him, he'd gladly oblige. They'd not discussed going to the mountains. She'd announced it.

Adam skipped across a hole in the street showing blacker in the black night, and lunged against the door into his room. When he flung himself into the shadowy light he found Edward in bed reading, and Jack with his pipe in his mouth pulling off his trousers. Adam giggled, staring first at Jack's bare bottom and then at Edward twisted uncomfortably under his paperback, the lamp set on the edge of his scuffed brown suitcase next to his head. Jack ignored Adam, continuing to undress in the silent room. Edward was lost in his thriller. Trapped, both of them, paralysed by niggardly habits of mind.

Adam's pity for them cleared his view wonderfully; the mean filthy room was exactly, he decided, what those two deserved. He wanted to go back out into the dark street, and leave them to stew in their blind cave without him. He'd stopped giggling. They were like the people at architecture school who couldn't stand colour, black and white only – even their clothes – or grey; like his mother whose clothes were beige or grey-green, the silly woman. Free himself he must. He was different, and something was happening to him now at last. He was in love . . . He'd always loved colour. He'd painted his bedroom poppy red when his mother was in France. Adam walked slowly over to his bed, because it was late and he was tired. He pulled off his bright blue sweater and stared at it for consolation in the stifling room.

Jenny and the mountains. Blues, golds, deep chestnut browns, a brash acrylic red sweater running down the tan street on a chubby dark-skinned little boy. He didn't mind that it was an artificial red, he wasn't prejudiced like so many purists. He had an open mind to all colour as long as it was lively and daring and exciting. Adam pulled his blanket up over him and shut his eyes. To feel himself alive was a delirium of colour . . . but it had always been there, his love of colour. Since the day he was born . . .

They were very high, but the grey crag of Mount Dikte was still way above. The village below looked like a pack of cards splayed haphazardly in the dark valley, condemned to dark afternoons by the mountains. What a flimsy, incidental place it looked, from their superior position in the sun, both remote and insecure. Adam hugged Jenny, imagining he could smell the sweet musk of wild sage in her hair. It was in the air, this strange scent of a golden afternoon, the tinkling of sheep bells and drone of bees completing the innocence and ecstasy of their spot. They'd taken the donkey road and climbed higher and higher on the rough stones, regardless of where it was taking them. Jenny had grown silent. Hugging her, Adam hoped to break into her thoughts and make her smile. She'd stopped smiling as soon as the view had opened out around them of tan hills and pink crags and, worlds away at the foot of their vision, the purply orange coastline.

'Please. I want you to smile. You look beautiful. Your hair is the same colour as the pines.' Adam ran his hand gently over her hair and kissed the tip of her nose.

'You're so young!'

'Why does that make you angry?'

Jenny broke away and stepped off the road, abruptly seating herself. As abruptly she jumped up, stung by a thistle. Further off she settled herself, giving a demure tug to her skirt. She clasped her knees and sniffed at the sage bush beside her, looking unhappy and uncomfortable.

'Don't be ridiculous. I'm not angry. I'm old. I wish I weren't.'

Adam kneeled down beside her, taking care not to kneel on a thistle, and started to feel her bosom. He laid his other hand on her knee.

'Why are you doing this when there's all this view to look at?' Finally Jenny looked into Adam's face, screwing up her eyes out of embarrassment.

'You need it,' he said.

Jenny pushed his hand off but continued to sit. 'I do not need it, thank you very much.'

'I need it then.'

'I don't believe you.'

'I do, I do, I do.' Adam had jumped up and was cavorting about in front of Jenny, pounding his chest. She burst out laughing. With a whoop of triumph Adam fell on top of her, pressing down her knees and pushing her over on to the sage bush. She was still laughing, giggling, and then tears ran down her cheeks. He kissed the tears and tasted the salt. He frowned, his sharp blue eyes shifting from her nose to her long narrow mouth, to the line of grey hair arching above her forehead. Jenny saw dismay spread across his face, his eyes wide and still. Carefully he climbed off her and Jenny sat up.

Just in time. A loud wheeze at their backs turned out to be a donkey carrying two tall tins of honey, the thick-set woman in a mustard scarf leading it peered at Jenny and Adam through the sage bushes. Jenny stared furiously at the view, afraid to show her face, while Adam slunk off, kicking aimlessly at stones. The donkey's hooves crunched forward a few steps, and the woman, in a rasping whine, asked Adam what he'd lost.

The impertinence of both the woman and the donkey (which let out a deafening ee-aw at this point) so irritated Jenny that she whirled around and showed her face. She shouted at the woman that everything was OK. But the woman's big, weathered face looked blankly back. With a demure tilt of her head the woman came forward, pouring forth in a low monotone incomprehensible Greek meant only for Jenny. It was the suspicious glance she flung over her shoulder in Adam's direction now that warned Jenny of the woman's intent. She took Jenny's arm and urged her to mount her donkey so that she could ride back to the village with her.

When at last the mustard scarf took in that Jenny was a bad woman and wished to be up in these mountains alone with the boy, it hissed at the donkey, yanking the lead; it hissed again, with eyes levelled on Jenny, before the donkey and it slowly withdrew from such foreign, evil carryings-on.

'They know we're here now.' Jenny sat hunched over her knees in despair. That woman had broken any courage left in her to dare and be shameless. Last night her courage was still strong. To tell Bill as he was undressing that she was off with Adam and see bewilderment raise his forehead in that familiar way was a pleasure. She'd surprised him. But the further they'd come through a dust-storm of road works, driving around the Japanese bulldozers which like shipwrecks had been left in the middle of the unfinished road, she'd begun to quail. They missed the turning to Symi and had to turn back. To lose her way frightened her and always made her want to give up. Adam noticed a sign on its side in the dust which had 'Symi' written on it. He urged them on, and she succumbed because it would have been shaming to go back after her scene. Slowly they bumped their way along, the bleached roots of pine trees jutting out of the cliffs above their heads. They reached the village. They drove on. They left the car under a tree beside a stream and began to walk. The dark, forbidding gorge had given way to wide skies and long views. Adam was obviously in a trance, exclaiming over the bright blue and the bright turquoise of the beehives. He hugged her, he kissed her ear, he ran his hand across her bottom. She walked on in silence, letting Adam go on as he liked, ignoring him as best she could. It was a new shame, a different shame from the shame of returning to Bill, that beset her now. The crags of Dikte glinting in the sunlight exhilarated Jenny. And then also made her wish she wasn't there with Adam. Better to be alone. She was demeaning all notions she ever had of how she should behave. A more daunting shame took a grip on her, and paralysed her longing for love.

The woman in the mustard scarf finished what the long view of tan hills tumbling down into the pink distance and the crags of Dikte rearing overhead had begun. Jenny felt herself undone and hopeless.

Jenny also imagined how her despair made her look even older and uglier, which would confuse and frighten someone young like Adam. She ground her chin into her knee.

Her hair pulled slightly. Quickly putting her hand there she felt another hand, Adam's; Adam had sneaked up behind her and was picking thistles out of her hair, which were very prickly and difficult to untangle. She whimpered.

'Don't worry,' said Adam. He gave a yank. 'There.' He left Jenny with both hands pressed to her head.

Adam faced Jenny now, his hands busy picking apart the thistle. He wasn't even looking at her, leaning his weight on one leg, propping his other foot on a stone.

'Thank you.'

Adam looked up and smiled. There was a thin line of damp hair down his temple where sweat made it curl. He was too hot in that blue sweater, his face flushed in dark blotches from the climb. But Jenny didn't tell him to take it off. She didn't want to sound like his mother. She rubbed her hands up and down her legs, thinking they'd better go back to the car.

'You were terrific,' Adam congratulated her. 'Really terrific the way you stood up to that woman.'

'She wanted to save me from you.' Jenny laughed.

'That's better.' Adam tossed away the thistle and threw himself on to the ground beside her. 'I thought you'd given up.'

'What?'

'Us.'

'I've known you since you were three months old. You were a very ugly baby.'

'I still am ugly.' Adam grinned up at Jenny, propped up on his elbows. She shook her head and touched the pale temple with her finger. 'Aren't I?' he persisted. Moved by the line of hot hair curling against his skin, Jenny found her voice wouldn't come when she replied to this insistent boy. She cleared her throat and shook her head, stretching her neck to compose herself.

'Still a baby?'

Adam laid his head in her lap. Jenny stared down at it, afraid to touch him.

'Someone will come to see what we're up to, you know,' she warned him.

'What?' Adam had mumbled something into her skirt which she couldn't hear. 'Get up,' she ordered him. 'Please.'

'I love you.'

'You what?'

'I want to make love to you.'

Jenny tried to push Adam off her, but both his hands clutched her thighs. He was amazingly strong, and determined. 'Don't be ridiculous,' she cried. 'Not here. You're too young. You can't. Think of your mother,' she babbled. He slid his narrow body up her, pressing her down on to the ground, kissing her, his feathery fingers groping with her knickers, on top of her. And she was responding, was giving in on this mountaintop where she had no business to be. Feverishly pulling at his trousers now, she helped. 'In the heat of the moment,' she croaked into his ear, trying to keep a shred of reserve, clinging to this strange, thin back, pressing him on to her. And all reserve went; in the heat of the moment the last shred was gone, while a little boy in an orange sweater with his older sister watched from the road, the long streamers of a red ribbon in the sister's dark hair crisp and still as they held on to each other's hands.

Above her the sinewy trunk of vine dangled gold leaves and purple grapes, only cracks of blue sky visible. Her host watched Laura spoon mouthfuls of the honey and walnuts his wife had just given her. He had a flat nose and wide mouth like a goat's. Laura wished he'd stop staring. Michaelis had just walked off. In a churlish fashion, as if Laura were suddenly an unwelcome appendage, he'd explained not to her but to this old man's wife where he was going. Laura chomped through the walnuts; the sweetness of the honey began to cloy. She held the sticky plate out from her, where it waited in her hand. No one would take it away.

Ridiculous. She'd agreed to come because Michaelis was so set on it, and it was a golden afternoon. But riding on the back of his motorbike through road works was unpleasant, and now where was he? She put her plate down on the ground and zipped up her blue anorak, shoving her hands into the pockets. There was the continuous mellifluous rush of a stream very near; she'd forget about Michaelis and enjoy this paradise of nuts and honey, falling leaves and water like a valley in Austria or the Lake District. Or the Cotswolds even.

A hand gripped her arm. It was her host. 'Where are you going?' he shouted. 'Stay!' He pointed to her chair.

Laura sat down again. She stared at her pink canvas shoes. She stuck her bluejeaned legs straight out. It was horrid of Michaelis, this. She was hating him, resenting now the lies she'd had to tell Christopher and Bill to cover up what she knew. On Thursday night by the fire up on the hill she'd suggested it was the Epimelete who had robbed the tomb, when Christopher said he wished he'd stop suspecting Andonis. She'd poured out more brandy, elaborating – all the time that she knew it was Vasilis and Michaelis. Damn Michaelis.

There was a slow heavy crunch of boots. Laura looked up and saw behind a shepherd in a brown headscarf a goat with long silky grey hair and curled horns. It was wild and beautiful. The shepherd tied it to a tree and dragged a chair over, sitting astride it to talk to the old man. He ignored her and the goat. The goat butted the tree, its sleek sides throbbing.

More men arrived. By now Laura felt almost as desperate as the goat. A fattish thirty-year-old in sandals and tight trousers collapsed on to a chair without a nod of his head. With him was a flattened face and pointed chin like a pear, with blue eyes so far apart they looked mad – this man moved closer to stare, his gnarled hands hanging at his side.

Laura found relief in a rough nail. She screwed up her mouth, and picked hard at it with her other hand. She tore it off which made the nail rougher, and was so distracted by this that she gnawed on it. The talk around her faded.

It was Christopher, bent forward holding his flask, his face flickering in the light from the fire, that needled Laura most. He would be up there again suffering another sleepless night. And she was doing this to him. More and more it bore in on her that Christopher was there because of her. She tore off another nail, feeling a right shit.

When Michaelis returned she had no nails left. She stuffed her maimed fingers into her anorak pockets. Michaelis introduced the bald, well-fed man with him who shook Laura's hand. He spoke English. He showed her his beautiful cane made of rosewood which he'd bought in London. He had been a general. He sat down beside Laura and described to her the beauty of these mountains where he came from and where he had returned to die. He looked too well to die, his white neck and jowls in his open-neck shirt still smooth. He could have been Laura's uncle who raised greyhounds on the Dingle peninsula. Or her father, she supposed. He too had retired to where his family came from, to his draughty cottage in Ireland.

'Don't you feel cut off?'

The ex-general smiled. 'Of course,' he agreed. 'My wife would agree with you certainly.'

'She lives here with you?'

'Why does that surprise you? She is my wife,' the ex-general reproved her, giving Laura his full attention. 'If you were my wife,' he told her, 'I should expect you to live with me.'

Laura laughed. It was wonderful to be talking English. Looking around her now, everything had changed. She was charmed by the faces surrounding her, by the crisp rustling of the gold leaves overhead, an idyll. Her host brought out more tables. His wife, in a flowery apron, had brought out plates of cucumber and olives and walnuts in their shells. The husband set out carafes of raki. Michaelis handed Laura a glass. How long ago? Had it been only ten days ago that Michaelis took her to his friend's still and sat cracking the nuts with a stone at her feet. Laura caught Michaelis's eye. '*Siyyia*!' She kept her eyes on him until she had taken her first sip.

'You speak Greek,' the ex-general praised her.

'Oh no.'

A spinach pie, a cheese pie, grilled pork, grilled lamb poured forth from the little kitchen. Cabbage salad, cheese, more olives. More tables were brought from the next house. Others arrived. An old man in a black polo-neck and a blood-red apron told how he'd been cut gullet to groin twice in eleven months and lived every day as his last. His wife came down the path shrieking that fifty sheep had broken into the vineyard. He waved her away. Michaelis started to sing.

The goat bucked and butted the tree. Its sides still throbbed with fright. Laura longed for someone to set it free.

Michaelis sang 'Philendem'. Their host stood up and raised his arms, his face washed of expression. He danced. Grapes came, and plates of cheese running with honey. A cool, clear night enveloped the party. The lone, grinding mutter of a motorbike approached and squealed to a stop. Raucous barks broke forth, and the mayor with his four brown hunting dogs joined them after a long day in his plastic greenhouse planting tomatoes. He, at least, had been working, thought Laura. Would he free the goat?

Three bare bulbs which dangled their light from the vine caught the curve of the goat's horns, its eyes and the knot of the rope. Michaelis's tousled hair and vulnerable face – in a khaki jacket and blue shirt – looked happy and proud. It was his party for her, she supposed. It was for this that he'd brought her to the mountains. Not for the scenery, the views, the air. Not for the reasons her kind of person would have come to the mountains. For company and song and food, too much food, he'd brought her. To show her off to his friends and perform before her. Had he noticed the goat?

'You look upset.' The ex-general sat back with his legs crossed, resting his hand on his rosewood cane.

'That poor goat!'

He smiled reproachfully. 'That is your Englishness. It is only a goat.'

'It doesn't know what's going to happen to it,' Laura exclaimed. By now its pain was Laura's. Its torture tortured her.

The ex-general waved his cane at the man in the blood-red apron. The blood-red apron stood up and motioned to the shepherd. Laura ducked. She knew what was going to happen. She hid in her hands.

The cry pierced the night. Laura's head was suddenly like a balloon bobbing aimlessly. She was going to be sick. She willed herself to focus on the iron leg of the table and took in deep breaths. Was the ex-general watching her, amused by her Englishness? The others were too busy drinking and singing and eating through all the food.

Tucked under his arm, wrapped in newspaper, was Michaelis's share, as he and Laura sped through the night back down to the sea. She'd seen the goat hung upside down from the tree with its throat slit. Everyone was given some. Michaelis was ecstatic with the success of their trip. He was still singing, Laura clinging on to him from behind. Laura rested her cheek on his shoulder. She smelt the sugariness of the wet newspaper and the mossy pungency of Michaelis's soft skin. She wanted him. With every bit of her she wanted him, dissolving into his back, hugging him as tight as she could.

Michaelis was singing 'Old MacDonald had a farm'. Laura joined in, grunting into Michaelis's ear her exuberant 'oink oink's.

There was no going back. The children had watched Jenny slowly disentangle herself from the thistles and sticky sage bush, picking the prickles off her white sweater. They'd seen her grey hair and white thighs. They wouldn't go away until Jenny told them, so thrilled were they and amazed. But Jenny was firm, surprising Adam by how she gently pushed them off down the road back to the village. As gently Jenny returned to Adam and smoothed back his damp hair with both her hands, kissing him lightly on the cheek. He hadn't expected this. She was so composed and so considerate of him all of a sudden. She hoped he hadn't scratched himself on the rocks. She looked closely at a scratch on his hand, tracing the red line with her finger.

'Shall we go on?' she asked.

'Where?'

'Or back?'

Strands of hair hung down the sides of her long face, endearing her to him with a fresh intensity, her present dishevelment his doing. Moved by his success, thrilled by her acquiescence, he threw his arms around her and crushed her to him, squeezing the tears from his eyes as he felt her soft body give in to him utterly. He was shaking, a confused, grateful boy, blinding himself with shut eyes as he clung on to her.

'Why are you crying?' Jenny tilted back his head and flicked away the tears with the back of her hand. 'I'm glad. Thank you.'

The ground around them purpled with the late afternoon, the rim of the sun now touching the tumble of hills stretching west. Very soon shadow would be on everything, the day nearly over.

'I came too quickly. I wasn't any good,' Adam burbled, stricken with the inadequacy of his performance on this momentous occasion. The place was momentous. And Bill's wife, his mother's friend, a woman he'd known all his life, as familiar to him as his parent's house, kissing him, loving him, utterly giving herself to him, suddenly, here on the mountainside, without a word of reproach or resentment or shame, was momentous. This was not anything he'd ever experienced before. This was new. And humbling. Had he bitten off more than he could chew? Ah, he found a reassuring certainty in that banal phrase.

Adam pulled back his shoulders and grinned, presenting himself, all of himself, with hands at his side, feet together, as Jenny's to do with what she liked.

It was an old hotel on the lagoon, without carpeting, with plain boards, iron beds. Their room had a balcony overlooking the small fishing boats and a row of white shopfronts. They were in the town where tourists came when they came to Crete. They'd gone on. Lying on their backs on the bed they watched a streak of light cross the ceiling, the chesty stuttering of a tourist's moped emptying the street below of other sounds.

Silence. A dull hum from the waterfront.

'Hungry?'

Jenny was enthralled by the silence. But Adam's words impelled her to raise herself on an elbow and look at him. She was too suffused in satisfaction to see very well what he looked like, now without clothes on, his elbows sticking out from under his head like clothes pegs. Such thin arms he had, the memory of his light body on top of her melting inside her, touching off a greed in her which was unseemly. She flopped on to her back, sobered slightly. They'd made enough love, she supposed. The smell of sage and the rough, taut undergrowth on that rocky slope came back to her, calmed her, excited her, distanced her suitably from Adam who was now on the bed beside her, who had food on his mind.

'Are you?' she asked.

'Are you happy?'

'Yes.'

'Was I better?'

'Of course you were better. Was I?' Jenny retorted, an impatient edge in her voice jolting Adam. He sat up.

'I suppose you're worrying about Bill. I'll telephone if you like and say the car's broken down. He'll blame it on me.'

Jenny regretted her impatience. Adam couldn't understand, he was too young, just a boy after all. So kind when he lifted her drooping breasts as if they delighted him – not put off as she was every time she looked in the mirror. She must be careful of the boy. She was, after all, responsible for him.

Bill could not be further from her mind. *She* was on her mind. This darling boy had spirited her away to another wonderful country, of what? Was it only sex? Was she a typical childless, middle-aged wife starved of sex? She wanted it to be more than that. She wanted to change.

'Bill's guarding the tomb. Let's eat.' Jenny had her hands behind her to fasten her bra. Her black pubic hair wasn't old like her other hair. Adam turned, so that she would not catch him looking. She might guess what he was thinking and be angry. He wanted her to keep as she was when she'd so surprised him on the mountain with her gratitude. He wanted her to love him, and trust him, and forget her age.

CHAPTER 28

It was Sunday night. Vasilis had been fuming ever since the Ephor left him in the middle of the road without a mention of a reward. 'Return the antiquities to *Kyrios Klystopha*.'

Why should he? He was the Greek. And he owned them now. He'd found them. What game was the Ephor playing? Whose stooge was he? Someone was on to him. Vasilis knew from the man's prim suit and natty handkerchief that he was scared.

He had tried to scare Vasilis. Vasilis was to save him from his persecutors. Vasilis paced his café; he knocked into the tables like a blind man. He railed at his wife for not sweeping the floor. He glared at anyone who entered the café. His anger was poisoning the air around him. Because he hated the Ephor for doing this to him. And he quite liked the Ephor, as well. He was obviously an intelligent man, a scholar and a Greek. No fool either. When he apologised to Vasilis for threatening him on his first visit, Vasilis recognised a shrewd man. It was clever of the Ephor, also, to have found out about his cousin. That showed he was resourceful, and desperate. He'd been right not to tell the English. He understood how that would unman Vasilis . . . and strengthen his resolve to sell and make money. His reputation was still intact. Damn the Ephor. The Ephor had pretended to believe in Vasilis, and like him. When he left he assumed that Vasilis would do as he had asked and return the antiquities to *Kyrios Klystopha*.

He was trapped. He needed the money. His hotel was costing thousands and he was in debt to the bank. He was the only man in the village to show initiative. Wasn't it right that he should flog antiquities to pay for it?

Vasilis grew more miserable as the days passed. If Michaelis had asked for his share it would have helped. But Michaelis avoided him. Then the old workman Andonis was summoned back by *Kyrios*

Klystopha. Vasilis had warned him to keep away, but he'd listened to the Englishman; he'd come back. At some point Andonis would tell. The workman seemed to trust the Englishman, and would expect the Englishman to keep him out of prison.

On Sunday night almost all the tables were taken. Families had driven out from Ierapetra, and Maria fed them pork chops and chips. Young men had ridden out on their motorbikes and were drinking and playing cards. Only on Sunday nights was there this much company. When Susan and Jack took one of the last tables and asked Vasilis for beers, he told them to wait because he was far too busy to look after the archaeologists who could come any time.

Michaelis walked in; Susan waved to him to join her and Jack at their table. Michaelis helped himself to three beers out of the refrigerator at the back and carried them to their table.

But it was only when the German strode through the door and gave a shout to the English that Vasilis's curiosity was aroused. The German took a chair from another table and dragged it over to join them. The German seldom came into his café. Vasilis watched them from behind the counter. He resented how Michaelis in the last few days had become the friend of the English.

'He stood in the doorway like this.' The German spread his legs and held out both arms. 'Like that!'

Vasilis came out from behind the counter with four beers.

'I'm interested. I love antiquities.'

'It's quite a museum,' put in Susan. 'Even a Cycladic import, a chance find.'

'That's what I mean!' the German persevered. 'I wanted to see this amazing museum. He blocked my way.'

Vasilis knocked the new beers against Michaelis's empties. 'Who?' he demanded.

Excited by Vasilis, Susan threw up her hands. 'The schoolmaster. He won't let him see his museum.'

Vasilis looked hard at the German, scornful of his clean pale shirt and big pale body. But there were Germans like Luther who'd stood up to the Pope, whom Vasilis admired. And there had been a German soldier in the war who became his friend and said Crete was the most beautiful place in the world.

Vasilis shrugged.

'But isn't he supposed to be setting an example?' Susan complained.

Vasilis smiled. 'He thinks I'm a Communist.'

In the doorway stood the schoolmaster in a starched white shirt, leaning on his cane. Vasilis backed away, the first to see him. He collected plates from the next table. Then Susan looked around and nudged the German, pointing her head at the door.

The German jumped up and shouted, 'Aha!' His 'aha' pierced the babble, the high guttural pitch strange. Heads turned; the schoolmaster in the doorway was obliged to react.

Like a bull the schoolmaster lowered his head and like a bull stopped short a few feet inside. He swung up his cane and threw back his head. Down came the cane, hitting the English table.

The babble stopped; only a child's whine was shushed. Vasilis pushed Maria back into the kitchen when she tried to come out.

The cane swept the bottles of beer off the table. The next instant it was flicked back to the schoolmaster's side. He stood immobile; the bottles made a dismal clunk against the wall.

'Murderers! Never forgive them. Never! Him' – the schoolmaster's cane tapped the top of the German's head – 'and him!' When it tapped Michaelis's head, Michaelis's astonishment at the touch of the cane on his red curls was ignored by the schoolmaster. 'Get rid of them!' The schoolmaster turned back towards the door, a rigid, tottering stick of a man.

The German grabbed the schoolmaster by the shoulders and threw him onto the ground. The old man and his cane sprawled between the chairs electrified the young men. Until then they'd watched smiling. But Sunday evening made them restless, the orthodox time for family meals out. There were eight of them. They threw down their cards, kicked over their chairs and made for the German. A mother shrieked. Michaelis pushed two of the boys out of the door. Maria screamed at Vasilis. Vasilis shook his head. Casually he tipped over the English table and waved them out. He waved out the boys. The German was in the grip of three of them, twisting awkwardly to free himself.

The Epimelete blocked Susan's way when she reached headquarters. She apologised when she bumped into him.

'They're killing that poor German. What do we do?'

'I'll go,' he told her. He walked off down the street, nodding to Jack who was just behind.

Susan expected to stun everyone. The violence of the boys and the sight of the schoolmaster prostrate frightened her still, so that she didn't take in that Christopher faced her in tears. Susan pushed on into the room and screamed, 'They're killing the German!'

Christopher pulled the door behind him.

'Are you sure?' asked Annabel pleasantly. She was helping Bill pass down bowls of soup. Jack stood behind Susan with his hands in his pockets; he watched everyone tuck into their soup. It was amusing. So English. Here inside headquarters it was supper time and a million miles from the brawl at Vasilis's a block away. Jack approved. He found his place. Thank God they'd escaped. Greeks were strange. Even the Epimelete had only nodded, although they were supposed to be friends.

'What should we do?' Susan was on her feet still. 'Bill!' she shouted.

'What?' Angrily Bill stopped ladling soup and glared down the table at her. 'Have some soup.'

Susan felt her face sting. Usually so scathing of others when they were upset, and finding herself now scorned, she plomped down into a chair and glued her eyes to her lap – she squeezed her thighs together, struggling to keep back the tears. She'd never seen such violence . . . When Annabel held a bowl out to her, she shook her head. When Laura burst into the room and exclaimed on the shrieks coming down the street, Susan did not look up.

Laura was full of energy from a lovely day sleeping on the beach. She helped herself to soup and sat down next to Bill. 'What a kerfuffle. From the shrieks and banging going on you'd think someone's being torn apart. Perhaps they're celebrating.' These casual remarks of Laura's brought silence, which Susan's screams had failed to do. Laura looked down the half-empty table. No Jenny, no Adam, no Christopher, no Edward. 'Where's Edward?' She was least concerned about Edward.

'At the cemetery.'

'Oh.' Laura gave Bill a guarded look; his ruffled appearance and noisy slurping of the soup didn't encourage her to ask more questions.

'The Epimelete's just been and announced he's arresting Andonis Markakis.'

'No!'

Bill ignored Laura's reaction.

Laura took her soup out to the kitchen and poured it down the drain. Annabel watched her in her red dress with the plunging neckline run down the room to the door. She was disappointed because there were only the four of them to admire her, thought Annabel.

Laura ran towards the commotion, the tight skirt of her dress impeding her as much as her sore ankle. She pulled up the dress so that she could move faster. In the street outside Vasilis's stood a crowd of men – it looked like the whole village. Vasilis stared out at everyone like Mephistopheles in his dirty black sweater, the sleeves pushed up to the elbow. Laura pulled her cardigan around her. Vasilis saw her and left his perch. He took her arm and walked her into the empty café where the tables were still littered with piled plates and crumpled napkins, one table on its side in a wash of beer and broken glass. But Vasilis couldn't explain to Laura very well what had happened. Only that Michaelis and the German had had a fight because of the colour of Michaelis's hair. This was all Laura could deduce from Vasilis's wild gestures, pulling at his hair like a madman and repeating Michaelis's name.

Laura left the café and the crowd outside and hurried to the end of the street where it was dark and empty. She picked her way over the rough ground in her sandals, making for the riverbed where Michaelis and the German had fled. Then she remembered the hut where Michaelis had taken her to see the still, up a back, narrow track behind the street, and wondered if they'd gone there. She retraced her steps, stumbling, cursing her bad ankle, afraid she'd turn it again. She stopped and listened. Where on earth was he? She was in a fever of determination now to own up and save Andonis. And Michaelis must know. Now. He too must own up, and report Vasilis.

Poor Christopher. She was sure that was why he wasn't at supper. He was by himself in his room, throwing back the brandies. She'd go to him and confess. She must.

But first Michaelis. Where the hell had he gone?

A curse on these high-strung Greeks who were always fighting. Not that she'd seen many fights, if any, but now, in her frustration, she imagined they were always at it. She was growing angrier because she couldn't find Michaelis, picking up her foot and rubbing the swollen ankle which had only just escaped a second spraining.

She'd go to his room. She'd wait there. Again she had to pass through the crowd of men who watched her flee in the opposite direction toward the sea.

The German sat on the bed with a towel wrapped round his head, Michaelis leaning over him with his back to the door when Laura squeezed through. The small lamp at the German's feet sputtered light up on to his chest and the white towel; the black figure in front ministering to him was unmistakably Michaelis. The long wide torso in short wide trousers dipped a cloth in a bowl and dabbed at the German's face.

'Michaelis?'

Michaelis hadn't heard Laura come in. He swung round to see who it was and kicked over the bowl. The German groaned.

'It's me.'

'Laura!'

Now she could make out a jug a foot away, near the lamp. Quickly she stooped down to refill the bowl with water. She stood close to Michaelis, holding the bowl for him. Dark stains on the pale shirt must be blood. More dark streaks on the German's face under the white towel looked ghastly in the bad light.

She whispered, 'What happened?'

'Sorry,' Michaelis apologised when the German started back, stung by the cold on his cuts. 'I can't heat the water. I have no fire.'

The big man sighed, a surprisingly compliant sound. Pitifully grateful, Laura thought.

'What did they do?' she asked, still in a whisper.

'They beat him up.'

'Who?'

Michaelis took the bowl from Laura. 'Go and get raki from Vasilis.'

Vasilis wasn't surprised. Nor curious. He handed Laura the raki, eyeing her plunging neckline and narrow bottom in the red dress as she ran back out into the street.

'There's a glass.' Michaelis pointed to a corner of the room where, as she felt around, her hand came to a tall glass. 'Pour!' She poured. 'Give me the bottle.' Soaking his cloth in more of the raki, Michaelis pressed it on the cuts. The German yelled and jerked back, putting up his arms. Michaelis pleaded. By now the smell of sweet raisins saturated the room.

'Give him to drink,' he told Laura. The German gulped down almost all the raki, and Michaelis finished it, handing the empty glass back to Laura. 'What about you?'

She refilled the glass and took a few sips. Michaelis finished bathing the German's cuts. He held him down with his other hand, ignoring the poor man's moans. Laura smelled the sweet smell of the drink mixed with the dust and darkness in this hovel of a room, when she remembered why she'd come. Fright gripped her. There was still Christopher! After this, after she'd warned Michaelis. Impatiently she took gulps of raki, cut off now from the drama here in the room, steeling herself. She wished now that the German would get off that bed and leave. When Michaelis asked her for another drink she didn't hear him. He held the lamp up to her red dress and whistled. She turned her head away. He set the lamp down and took the bottle and glass from her.

The German fell against the door. Michaelis took his arm and eased him out into the night, turning to Laura whom he ordered to wait there. Reluctantly Laura acknowledged that the German was too drunk to get himself to bed. She sat herself down on the bed hearing their feet scrape and trip outside, before there was silence.

'You care. Why?'

'I can't let Andonis go to prison. I've caused damage enough.'

'What damage?'

'I've deceived them. Poor Christopher's exhausted from guarding the tomb night after night. And now Andonis. It's too much.'

'He's a fool.'

Laura was startled. Michaelis stood in front of her with legs apart and hands on hips like an inquisitor. She was still on the bed, with her hands tucked under her.

'You're wrong,' she whispered, sounding more hesitant than she'd meant.

'I think you love him. Doesn't he realise? Your Christopher can't protect the old workman. He's a foreigner.'

Laura slid off the bed and pulled down her skirt. Dark holes in Michaelis's face were devouring her pale skin, which was amber in the faint light. The shadow between her breasts exposed her. If only she hadn't worn this dress. She crossed her arms.

'Only now I learned about Andonis,' she defended herself, catching hold of her woollie and pulling it close round her. A spasm of shuddering threw her sideways, so that she fell against the wall. Michaelis steadied her.

'What are you afraid of?' he asked when she pushed him away. He laid his hand on her arm.

Laura shook her head, more wisps of hair falling from her bun at the top.

'You're lovely,' he sighed. Both hands now hung limp at his sides.

'Goodbye.' Laura took the initiative. She extended her hand to him, keeping well away.

Laura knocked, relieved to find Christopher's light still on. But when she walked in he froze, the book he held stopping in midair, he on his back beneath it. Slowly he raised himself on an elbow and stared. Was it her dress? He frowned with eyes wide.

'You'd already left or I shouldn't have come down here. Sorry!' she began, angry that she was so afraid of this handsome man. Damn it!

It was then she noticed his thin white thighs on top of the crumpled sleeping bag. She felt her courage return – wondering why he didn't feel cold. 'I need to have a word with you.' She waited for him to offer her a chair.

Christopher sat up and leaned his back against the wall. 'It's awfully late,' he sighed, irritated. Couldn't she have waited until the following day? Her dress was too much at this hour.

'May I sit down?'

He pushed the paperbacks off the chair. His nerves were still in a bad way. He shut his eyes to make at least the sight of Laura go away. He was so tired. He moaned, rubbing his face with his hands.

'Are you all right?'

'Yes.'

'Just tired?'

'Yes.'

'They're arresting Andonis. Bill told me.'

Christopher nodded. He pulled the sleeping bag right up to his chin, and with his shut eyes tried to concentrate on a plan of work for the next day. He would not lose his temper. Nor burst into tears. He'd already cried twice this last week. Was it natural? Or was he on the brink of a crack-up? God, he hoped not! Whatever the truth, he'd try and hide it from Laura. 'You'll help Edward and Michaelis tomorrow. Bill needs Manolis,' he rapped out.

Insulted by those shut eyes and his failure to offer her a drink, Laura asked if Christopher had any brandy, knowing perfectly well that he had plenty. He fished under his pillow and handed her the flask; their eyes met, a wariness and reluctance on both sides caught off balance for a brief moment, which quickly became a memory as both looked down at their hands and remembered how they liked each other. And it was on the heels of this that a wave of dread winded Laura. She squeezed the flask. What would Christopher do? Laura gulped brandy and coughed, the flask clasped to her as she coughed her guts out – as it felt. She couldn't bear to be hated. As a child, she'd appease even her worst teachers like the mean Miss Workman, needing the approval of absolutely everyone. Afraid, otherwise, that hate of her would spread and isolate her, and that for the rest of her life she'd be alone and unloved, which would be hell.

When she told her mother that her father had gone to Ireland her mother blamed her. She'd never stopped blaming her for what she hadn't done. Laura had been suffering her hate ever since, nullified by her mother's wan face, although she could see how it would soften the shock of being abandoned. She still tried to change her mother's mind. They met for lunch. She'd stay to keep her mother company. She'd take her mother shopping. She remembered her birthday. Every Christmas she gave her a large bottle of Chanel Number 5 although her mother never said thank you. She expected it, but was never grateful. And yet Laura would not give up until her mother died. She must hope that her mother would finally forgive her.

Laura handed the flask back to Christopher. She was flushed from coughing. Christopher wondered why on earth she had put on that dress. For him? Was it to vamp him? He smirked and took a nip of the brandy himself before he screwed the top on and laid the flask carefully down on the floor beside his heap of paperbacks.

There was no wind. The night was clear and cold. No sounds

penetrated the murky room and their silence as Laura sat up straight in the rush-bottomed chair and stared at the lamp hanging on its nail above Christopher's sleek hair.

'I'm keeping you.' As she often did when embarrassed, Laura turned up her dress and examined the hem for loose threads. The poor light made it impossible to see, but she picked at an imaginary loose end. 'I've come to own up.'

Christopher burst out laughing. He couldn't believe she was as nervous as she pretended.

'Don't!' Laura screamed.

Flummoxed, Christopher fumbled for the flask. Laura leaned over and shoved it at him. 'It's not a pleasant thing to do at my age,' she rushed on. 'I've known for some time who robbed the tomb, but I haven't told you because I didn't want you to know why I know.' She glared at Christopher's symmetrical eyebrows over the wide-apart eyes, stiffening herself for his reaction.

'Don't be angry. Please,' she begged.

Christopher was still faintly amused, and wished she'd get to the point. 'Of course not,' he reassured her.

'Vasilis and Michaelis did it. Vasilis organised it. Andonis they only pressed into helping because he was there. That's why I've come now, when I heard they were going to arrest him.'

Laura gripped the edge of her chair, dreading the inevitable question. Her knees showed. Slowly she eased off one hand and smoothed back her skirt, not daring to take her eyes off the floor.

'How long have you known?'

She looked up. 'How long? Don't you want to know why? Since Thursday is how long I've known.' The great emphasis she put on 'long' relieved her shame, Christopher's uncanny calm giving her this chance to scold him.

'I wish you'd told me sooner.'

Laura's cold hands gave little comfort, pressed against her hot cheeks. She sat very still and took light soundless little breaths.

'Why do you care what I think of you?'

Now Laura was scared witless. Christopher's face became a blur; the room paled, dizziness nearly toppling her off her chair. She rocked back and forth and swung her legs. 'Do I?'

'Or you would have told me on Thursday.'

'Would I?'

'Wouldn't you?'

'I don't know if I would or wouldn't have.'

Christopher threw off the sleeping bag. He turned the backs of his thin legs to Laura while he pulled on his trousers. His heavy shoes scraped the floor; he dropped them before he started slowly to put them on. Laura's fear made him afraid. They must leave this room. That's all he could think to do. Turn on the Camping Gaz next door and discuss what he should do now. Was she afraid for Michaelis's sake?

Christopher sat for a long moment bent over the tied laces – seeing the figure behind him on the chair in the red décolleté dress. Although she was so afraid, she'd come. It brought back his fright when Bill had held out the receiver to him. He'd nearly fainted, he had been so ashamed of what he'd let happen.

'Laura?' Christopher peered round. Wisps of her unruly hair fringed her wide face. She was squinting at him as if she expected him to throw his shoe at her. 'You are very sweet. Don't . . . I love you, Laura.'

Laura stopped rocking. 'What did you say?' She pushed the hair out of her face. 'Oh, for God's sake!'

Christopher walked over to her. He lifted her hands off her lap and kissed her forehead, her hands still in his. He lifted her off the chair until they stood face to face. Laura ran her hands up and down her arms. 'This is ridiculous. I've done my confession and I'm sorry. Isn't that enough? I know I've been awful. The other night up there on the hill with you and Bill, making up that story about the Epimelete . . . You can't forgive that! Bill made me go. But it was horrible of me. I was pretending I didn't know anything, when I knew everything.' She threw back her head, chin up, eyes aglare, with such haughty courage that it was funny. Christopher grinned.

Laura slapped his face.

Christopher burst out laughing.

Aghast, Laura collapsed on to the chair and buried her head in her hands, indignant that anyone could be so cruel. When she let him lift up her head she stared dully up at his long neat face, threatening her with compassion.

'I forgive you.' There were tears in his eyes.

He forgave her! The words bounced off her self-contempt like a ping-pong ball, his fatuous intention of forgiving her jiggering off in a corner. What an extraordinary thing to say in this crude little room dropping mortar on to the earth floor like clinker, just after she'd confessed to deceiving him and sleeping with one of the robbers! Was she that unimportant? Was it the room which had lost him any sense of common civility? At least he could have pretended she'd done him wrong. Or was he too arrogant? Christopher was still holding her head in his hands, the wretch.

'I don't understand you,' she mumbled.

Christopher let go of her head. 'I hate to see you afraid. It's how I must have been looking.' He studied his shoes. 'I've loved you since you fell off the lorry. I'm saying I love you. I love you.'

'And what's that supposed to mean?' Laura snapped.

Suddenly Christopher yanked Laura off the chair and pushed her into the next room. He forced her to sit down on Jack's stool. Fumbling for matches, grunting from the strain as he leaned across Jack's drawing board to reach the Camping Gaz, he lifted it off its nail and pushed back the lid. Its glare and sizzle broke over the room. Laura didn't think he looked at all afraid.

Laura's spirits were picking up. Perhaps it was Christopher's temper. She'd seen plenty of that temper, but this time it relieved her. She crossed her legs and made space for her elbow among Adam's clutter. The banal assortment of pencils, tapes and rulers and the red and white ranging rods stacked in the corner comforted her. Christopher went back into his bedroom for the brandy.

He frowned at her from the doorway, forgetting the flask in his hand. Laura smiled, waving to him to pour out the drink: she would show him how she was no longer upset, how he could trust her now to be herself, how she would not slap him on the face ever again. (The red slash on his chin embarrassed her.) 'Anyway,' she volunteered, smoothing down her dress and breathing in to relieve the tightness of the waistband. 'We've probably said all there is to be said. You're a good sport, and I'm jolly grateful. So let's drink to the fall of Vasilis and the recovery of all those goodies. They must be somewhere.'

Christopher hadn't budged. He still frowned. He was not pouring out drink. He stood with both his feet together, taking no notice.

How could he make Laura love him back? She must. But she wasn't believing him. The race was on; he'd left the gate, blurting out that he loved her. How to go on from here? She was hard. She gave nothing away – only that adorable dress. Whatever would his mother have said about it! In London at parties he was always frivolous; it was the way he preferred to be usually. It was so easy. But it would lose him Laura now and he might not have another chance. He was too old to wait.

'No.' His voice was low and so serious that Laura flinched. 'Please.'

Laura bit her lip and shivered, close to giggles. But she mustn't. It was ludicrous. To be loved by this old bachelor? How she'd longed to be loved. So many dreams of how it might be, and never was. By now she was too cynical to believe even an awkward lunger like Christopher. He scared her though. This was not what she was used to. Such earnestness! She liked it.

Should she let him go on? Should she risk it? No. He'd caught her off guard. Love wasn't risk. What she needed was home. Christopher was in a muddle. Everything here was muddle – no place for love. Laura picked up a tape, squinted at it and dropped it back. But he'd surprised her. And how!

'I really don't know what you're talking about,' she mumbled, looking serious back as she ran fingers down her cheek.

Christopher grinned. It made his ears stick out from under his brushed-back hair. Or had she never noticed his ears? The tip of his nose pointed at her. Such impertinence – his grin was as knowing as Michaelis's whistle. Laura squirmed. How had she missed this devil in him? Too scared. And he was too scared of her! He wasn't now, though.

She grinned back.

He threw down the flask and pulled her off the stool, hugging her so hard that Laura coughed and begged him to let her breathe.

Laura leaned back against his arms. His face was delightful. They both laughed. Christopher held her as she burrowed her forehead into his shoulder and rubbed her wet cheek against his expensive shirt.

Only a small paraffin lamp burned on the counter, left there by Maria. Vasilis was hunched over a table asleep. Michaelis knocked Vasilis's head on the tabletop and pulled it right back hard by the hair. Vasilis yelped and sputtered, wiping drool off his chin with the back of his hand.

'What d'you want?' he grumbled.

'Wake up!'

'Go away. It's too late.'

'No.'

There were still bottles on the floor and heaps of dirty plates littering the tables, one table on its side, chairs chaotically overturned.

'Why hasn't Maria cleared this up?' Michaelis asked, amazed by the mess.

Vasilis blinked and groaned, rubbing his eyes.

Michaelis kicked away a bottle which skidded off at a tangent, and dumped himself down on the nearest chair. 'I'm telling the police,' he announced. 'Wake up and listen! Did you hear me? I'm telling them first thing in the morning.'

'You're crazy!' Vasilis shouted, instantly awake, furious. 'What d'you come here and tell me that for? Fool! Idiot! Bastard!'

Michaelis leaned back in his chair and stretched out his legs, resting his arm on the table.

'You'll go to prison,' Vasilis warned him. 'You'll be eaten up by rats and fed dog shit.'

'So will you.'

'Ah, but I've got a cousin. You haven't.'

Michaelis shrugged, not bothering to answer. Vasilis grew uneasy. His low forehead and black brows were pulled together, puckering his whole face into a anxious frown. In a much lower voice he carried on, leaning up close to his stubborn accomplice. 'What's wrong? Is it the German? Threatening you?'

Michaelis shook his head. 'What are you going to do? Tomorrow morning I tell them.'

Vasilis argued, but Michaelis was like a mule, obstinate as Hades. He wouldn't listen to reason. Vasilis got to his feet and shook him by the shoulders but still he wouldn't change his mind. It was no use. Quicker than it might have appeared to Michaelis, Vasilis acknowledged to himself the corner he was in, his cunning devising a plan while he still argued that Michaelis was about to do the stupidest thing of his life.

Vasilis did not tell Michaelis that the Ephor suspected him already. He did not want Michaelis ever to know that. Rather, when he at last gave up, letting his head drop and his eyes close, he begged Michaelis to wait just twenty-four hours. No longer. That was long enough, he said, for him to arrange things; then Michaelis could tell the police, if he still felt that that was what he must do.

CHAPTER 29

The Epimelete found the police officer at his table in the station reading the newspaper. The officer was alone; he greeted Costas with a casual 'good evening' as he continued to read the paper. Which Costas minded, not liking to feel uneasy before a man he instinctively despised. Costas pulled up a chair and dumped his car keys on the table; the loud clunk of keys forced the officer to look up.

'So? Why have you come?' asked the officer.

'To report the robber.'

'You know who it is?'

'Yes.'

'How do you know?'

'I have found out that he was the night watchman.'

'The old man.'

Costas bowed his head. The officer raised his eyebrows.

'He's no robber. He's too old.' The officer smoothed the folds of his paper with flat hands.

'But he was there. He's an accomplice.'

The officer shrugged.

'Arrest him. He will tell you who the others were. He's afraid. He knows too much.'

'And the antiquities? Have you found them?'

Costas pointed at the table. 'He'll know. Ask him.'

The officer stood up. 'Should I ask the Ephor if he agrees?'

Costas also stood up and jingled his keys. 'I shall tell him. I go back tomorrow to the museum. The antiquities must be found before they leave the country,' Costas warned the sceptical officer.

Costas hurried out of the door and down the steps, zipping up his leather jacket. When he reached the street he broke into a run.

Sotiris Georgakis had been unhappy for two days. He couldn't bear to suffer for longer. But when he told his cousin what he was going to do, his cousin cackled like a rooster and teased him that he wouldn't dare. Disgusted, Georgakis put on his cap and rode back into town. He gave up his cousin's supper. Instead he roamed the streets behind the market, puttering up one and down another until he'd calmed down. Tomorrow he would ride out to the village for the last time, he decided, whatever his cousin thought. His motorbike would carry him like the white horse of St George.

Georgakis accelerated out on to the waterfront where his people walked and talked on Sunday night, happier now. Tears filled his eyes as he fell in, and felt himself on the brink of a great action. The phoenix rising from the embers, an avenging angel for his government of honest, clean, faithful men, he would be all of this. He respected his people. He respected Papadopoulos. He respected God. He skipped as he pushed along the motorbike, his spirits released finally from two glum days.

That young Englishman was scum.

How long they could keep their affair a secret, Adam did not know. Or care very much, although Bill face to face was a frightening prospect. Would Bill pretend he suspected nothing? That would be easier. The dig must go on, after all. Adam relished Bill's come-uppance. Finally Jenny would show the man that she existed. Oh, how richly Bill deserved this betrayal from the wife he'd ignored.

Too elated to sleep, Adam walked. Jenny would not walk with him, anxious to make sure Bill was all right when they finally made it back. It was too late to eat. But they'd had an enormous lunch, and he couldn't, anyway, have faced the others at supper.

The others would be like a wet winter day in England; they would assume he was as low as they. But he wasn't. He was happy. He was incredibly, wonderfully, triumphantly happy. Everything made sense, and there was possibility everywhere now. He was lucky. He was strong. He was able. Give up architecture, he would. Become an archaeologist. Why not? It was a much more interesting life. It was a challenge for him to reconstruct old buildings. He was good at it, he thought, convinced that the Minoans were much more civilised than people now with their plastics and tower blocks and supermarkets. The Minoans built wonderful houses and courtyards and village streets out of beautiful local stone, in places they loved, treasuring their views. Adam thought how people needed to be told that we were backward now, and crude, and ignorant compared with civilisations thousands of years ago. It would be his mission to show that we made deserts of our lives, turning our backs on the true and the beautiful and the good, *and* the right and the possible, like blind lemmings. Poor everyone. Adam was overjoyed.

Growing tired, Adam turned back. He'd passed the police station. He passed it again now, keeping to the far, darker side of the street. The Epimelete he recognised, rushing down the steps in his important leather jacket and black glasses. He called to Costas on an impulse to be friendly. Costas stopped. Adam caught up with him and for a short time they walked together, until Costas rushed off to the left, wishing Adam good night.

Adam slept. Jenny did not. Jenny was in the grip of a sickening, aching lust which gnawed at her. She turned on to her side, feeling the hard canvas of the camp bed numb her hip. On to her stomach she turned for relief, stretching the terrible longing in her belly for that willowy, slithering body of the boy who'd satisfied her miraculously, exhausting her with his demonic energy for love. Their bodies all Saturday night and Sunday morning had twisted and turned and slid about on each other, the memory now of their long steady climaxes of love on the creaking hotel bed a nightmare of loss which kept Jenny awake most of the night. Bill snored beside her, exhausted from his sleepless Saturday night, guarding the tomb for Christopher. He was already asleep when Jenny returned. His long shape and woolly head jutting out of the top of the sleeping bag didn't move as she tiptoed about the room, rummaging under piles of things for her nightie.

Nor did she meet Bill in the morning. He was already gone when she opened her eyes and saw the jagged line of daylight around their door.

'It'll be salami and salad from now on. I can't go on. It's no use,' Margaret announced, blocking Jenny's way when Jenny reached headquarters. 'I'd made carrot soup especially because it's Adam's favourite.'

'We had trouble with the car.'

‘I quit anyway. I’ve had enough.’ Margaret flounced backward in her clogs and sprawled herself over a chair, watching Jenny hurry through to the kitchen to make coffee. ‘What are you going to do now?’ Margaret called.

Jenny stood with her cup of coffee in her old brown skirt and guernsey, warming her hands on the cup as she steeled herself for a session with Margaret.

‘We must go in a minute.’

‘Where?’

‘Have you made a list?’

Margaret shook her head. ‘Mary Elizabeth didn’t turn up either.’

‘But she never eats much. Christopher?’ Margaret nodded. ‘Annabel?’

Margaret stuck her bare legs out into the room.

‘Is there any carrot soup left?’

Margaret shrugged. Jenny went back into the kitchen and found that there was very little left in the pan. She poured the rest of her coffee down the drain, pulled their shopping baskets out from under the trestle table and ordered Margaret to find a pencil and paper. They’d make the list in the car.

‘Why did you go off with Adam?’ Margaret sniffled as she followed Jenny to the car. ‘Bill looked a wreck.’

Jenny threw the baskets into the back. She raced the engine to rid it of hiccups and started off across the square. ‘Is your door locked?’ she asked. Margaret was slumped in the seat next to her with those bare legs which made Jenny shiver straddling their two bulky handbags. ‘Let’s see how quick we can be. I don’t want to be long.’ The car lurched off the cement street into the ruts made by tractors and the rains. They bumped across the gravelly riverbed. Up the other side Jenny picked up speed, nearly running into a motorbike coming the other way around the blind curve. She swerved and stopped, gearing down to first.

‘It’s that policeman!’ Margaret pulled her head back in through the window. ‘The one you hate, who interrogated us.’

Jenny checked in the rear-view mirror but he was already out of sight, only a puff of dust showing where he’d passed.

Bill ran the last bit past the Merry Widow café and the kiosk, but Jenny had already left. It was only eight o’clock. Annoyed and disappointed, wishing to have a word with Jenny and see how she was, as well as to suggest Adam go in with her to buy graph paper, he roamed the dark kitchen and found the leftover carrot soup. Since Adam was on his last piece of graph paper, he supposed he should go over to the cemetery hill to help Christopher, since there was no way now to buy more paper until the evening. He sipped the soup, gulped down the rest and started back, hurrying so that Adam wouldn't leave the site before he got there.

Adam also was disappointed, irritated with Bill for not leaving Jenny a note. He forgot that he'd not told Bill he needed more paper until an hour ago. He would punish Bill by crossing the valley at its widest point, and perhaps even stopping to sketch. He trudged off in his blue sweater and purple jeans, in a mood – a mood which Bill for once thought was justified. Why had Jenny been in such a hurry? Bill felt lost and afraid. He could not afford to consider what Jenny might have done. His suspicions sickened him. He would carry on believing she'd done nothing. Otherwise . . . he didn't know!

It was a clear day, chilly but sunny, with faint streaks of cloud piling up into thicker cover over the mountains where it was probably raining. The close, dusty warmth in amongst the orange and lemon trees blotted out the sky and the chill; grit from the rough clods of ploughed earth slipped inside Adam's socks. Twice he stopped to unlace his black shoes. But he didn't mind. He was enjoying his walk. To escape the others was again what he wanted; the fruity smell was delightful.

Up the other side he climbed, suddenly coming upon the old village. Hadn't Jenny and Laura walked here the morning there was such a wind? It was an extraordinary spot; bare crags loomed over the low dry-stone walls of collapsed houses and a street. The straight white branches of a fig tree choked a deep well like arms reaching for sunlight. Adam was thrilled by the aura of a place so recently abandoned. The proud branches of the fig had survived abandonment in the clear light and dazzled Adam. They made him want to draw. He settled on a wall and pulled his small pad out from his pocket.

He was so lost that he didn't hear the crunch of footsteps. He jumped when the German hollered his name. The tall blue figure just in front of him angered Adam instantly, a wretched shadow on Adam's pad blocking the light. Vasilis was beside him. 'You come too. Come with me!' Vasilis ordered. Adam shook his head. Vasilis waved his hands in front of Adam's nose, insisting. But still Adam refused.

'He wants you to come. He has something to show,' the German explained in his guttural English. 'Are you drawing something?'

'Drawing something,' Adam muttered. He scowled at the two interfering men. 'I'll come in a minute, when I've finished.' He watched them move off towards a house that was very white, with an arched doorway flanked by strings of onions and a black frying pan. 'I won't be long,' he called, suddenly curious.

With the small pad on his knees again, he returned to his drawing. He heard more footsteps but ignored them, eager now to finish. He was pleased. He held out the pad, his crossing lines covering the paper with his image of the tree. His eyes had shifted to the tree, were on the real thing, on the sleek white branches fanning out into the sunlight, when – a crack and something hit his back . . . he whimpered as he slid across the branches, hitting the ground, a horrible ache twisting him, until it faded into the faint buzz of a motorbike.

In the dark, low-ceilinged room just inside the archway the German could see three large sacks through a fretwork of rusty springs, the sad-looking bed up against the wall. Vasilis grabbed one of the sacks and pulled it into the middle of the room, ordering the German to pull out the others. 'You look and then you take it to the English.' Vasilis told him, waving to the German to hurry up. The sacks were heavy and awkward. The German felt in the first one and came on something smooth in among some twigs. Carefully he freed it from the twigs, something, an arm . . . catching on the sacking . . . one shot, suddenly, dulled by the small room. Very close. And silence then. Until there was the sound of a motorbike passing below. The German ran out through the archway, the shot was so close. The trophy of a bare-bosomed woman waving a snake was slung over his shoulder as he burst into the daylight.

Laura was keeping Christopher's notebook. She watched from the edge of the trench while Christopher picked with his knife at a patch of loose earth in the bedrock. They'd reached bedrock at last. Christopher gave it a last look, on his knees at the bottom of the chamber. Laura told him to give up and get out. She was writing down that Christopher thought a round patch of loose earth might hold something more but was proved wrong. She shut the notebook and leaned on it. Christopher climbed out and brushed off his trousers.

'So that's done,' he muttered.

'Glad?'

'Very.' He took the notebook.

'Don't you believe I can do it?'

He flushed and bowed his head. Laura laughed.

'How distrustful of you!'

'But soon you will!' He grinned.

Laura turned. A shot. Its suddenness surprised her. It rang round the bare hills like a cracked bell, the sound followed by no other, living on for another second before it died. Strange. Christopher caught the dismay on Laura's face. 'Odd,' he agreed, looking up into an empty sky.

'Some poor sparrow's had it,' remarked Susan, seeing no reason to be dismayed, or surprised either.